Focus in High School Mathematics: Reasoning and Sense Making in Geometry

by

Sharon M. McCrone
University of New Hampshire
Durham, New Hampshire

James King
University of Washington
Seattle, Washington

Yuria Orihuela
Miami Dade College
Miami, Florida

Eric Robinson
Ithaca College
Ithaca, New York

NATIONAL COUNCIL OF
TEACHERS OF MATHEMATICS

Copyright © 2010 by
The National Council of Teachers of Mathematics, Inc.
1906 Association Drive, Reston, VA 20191-1502
(703) 620-9840; (800) 235-7566; www.nctm.org
All rights reserved
Second printing 2011

Library of Congress Cataloging-in-Publication Data

Focus in high school mathematics : reasoning and sense making in
geometry / by Sharon M. McCrone ... [et al.].
   p. cm.
 Includes bibliographical references and index.
 ISBN 978-0-87353-641-7 (alk. paper)
 1. Geometry--Study and teaching (Secondary) 2. Reasoning--Study and
teaching (Secondary) I. McCrone, Sharon.
 QA461.F63 2010
 516.0071'2--dc22
                           2010027427

The National Council of Teachers of Mathematics is a public voice of mathematics
education, supporting teachers to ensure equitable mathematics learning of the
highest quality for all students through vision, leadership, professional
development, and research.

Printed in the United States of America

Table of Contents

Preface

Focus in High School Mathematics: Reasoning and Sense Making (NCTM 2009) captures the direction for high school mathematics for students in the twenty-first century:

> Reasoning and sense making should occur in every mathematics classroom every day. In such an environment, teachers and students ask and answer such questions as "What's going on here?" and "Why do you think that?" Addressing reasoning and sense making does not need to be an extra burden for teachers struggling with students who are having a difficult time just learning the procedures. On the contrary, the structure that reasoning brings forms a vital support for understanding and continued learning. Currently, many students have difficulty because they find mathematics meaningless.... With purposeful attention and planning, teachers can hold all students in every high school mathematics classroom accountable for personally engaging in reasoning and sense making, and thus lead students to experience reasoning for themselves rather than merely observe it. (NCTM 2009, pp. 5–6)

This new publication urges a refocusing of the high school mathematics curriculum on reasoning and sense making, building on the guidelines for teaching and learning mathematics advocated by NCTM in *Principles and Standards for School Mathematics* (NCTM 2000). *Focus in High School Mathematics: Reasoning and Sense Making* makes the case that reasoning and sense making must reside at the core of all mathematics learning and instruction, at all grades. Moving forward from *Curriculum Focal Points for Prekindergarten through Grade 8 Mathematics* (NCTM 2006), *Focus in High School Mathematics: Reasoning and Sense Making* also addresses the need for the continuation of a coherent and well-articulated mathematics curriculum at the high school level.

The underlying principles of *Focus in High School Mathematics: Reasoning and Sense Making* are "reasoning habits" that should develop across the curriculum, along with "key elements" organized around five content strands. The book provides a group of examples that illustrate how these principles might play out in the classroom. Historically, NCTM has provided supplementary materials to accompany major publications that present official positions of the Council (e.g., the Teaching with Curriculum Focal Points series for *Curriculum Focal Points for Prekindergarten through Grade 8 Mathematics,* the Navigations Series for *Principles and Standards for School Mathematics*, the Addenda Series for *Curriculum and Evaluation Standards for School Mathematics* [NCTM 1989]). In keeping with this tradition, a series of supplementary books, Focus in High School Mathematics, provides additional guidance for ensuring that reasoning and sense making are part of the mathematics experiences of all high school students every day.

This series is intended for secondary mathematics teachers, curriculum specialists, mathematics supervisors, district administrators, and mathematics teacher educators. *Focus in High School Mathematics: Reasoning and Sense Making* underscores the critical role of the Process Standards outlined in *Principles and Standards* and provides a foundation for achieving the principal goals for the mathematical experiences of all secondary school students. Each volume in the Focus in High School Mathematics series presents detailed examples of worthwhile mathematical tasks, along with follow-up discussion. The examples and discussions are intended to help classroom teachers understand what it means to promote sense making and to find ways to increase it in their classrooms. The material could also be used as classroom cases in professional development. In addition, supervisors, curriculum specialists, and administrators might use the examples and discussions to catalyze conversations about shifts in the high school mathematics curriculum to bring them into alignment with the goals of *Focus in High School Mathematics: Reasoning and Sense Making.*

Although the books in the series focus on a particular content strand or principle of school mathematics identified in *Principles and Standards,* they are not intended to outline a curriculum for a particular area or topic. In fact, many of the examples in the books point to potential connections across content areas and ideas.

The authors of the present volume, *Focus in High School Mathematics: Reasoning and Sense Making in Geometry,* extend thanks to the following reviewers for providing valuable feedback and support during the development of the book:

John A. Dossey, Illinois State University (emeritus), Normal, Illinois
J. Michael Shaughnessy, Portland State University, Portland, Oregon
Neil Portnoy, University of New Hampshire, Durham, New Hampshire

In addition, the authors express gratitude for the guidance and assistance that they received from W. Gary Martin, Auburn University, Auburn, Alabama, who served as the chair of the writing group for *Focus in High School Mathematics: Reasoning and Sense Making.* The authors are also indebted to Gary for a draft of the stepped-pyramid activity that they used in chapter 3 (in the section Developing a Formula for Volume). Finally, Eric Robinson's coauthors give special thanks to Eric for his service as a liaison to them from that writing group.

General Introduction to the Focus in High School Mathematics Series

Focus in High School Mathematics: Reasoning and Sense Making addresses the need for reasoning to play a larger role in high school mathematics:

> A focus on reasoning and sense making, when developed in the context of strong content, will ensure that students can accurately carry out mathematical procedures, understand why those procedures work, and know how they might be used and their results interpreted…. Such a focus on reasoning and sense making will produce citizens who make informed and reasoned decisions, including quantitatively sophisticated choices about their personal finances, about which public policies deserve their support, and about which insurance or health plans to select. It will also produce workers who can satisfy the increased mathematical needs in professional areas ranging from health care to small business to digital technology. (NCTM 2009, p. 3)

Focus in High School Mathematics: Reasoning and Sense Making provides an outline for how reasoning and sense making might play out in core topic areas of the high school curriculum: numbers and measurement, algebra, geometry, and statistics and probability. The topics and examples contained in this publication and the supporting volumes do not represent an exhaustive list of topics that should be covered in any particular course or curriculum. The examples are meant to illustrate reasoning habits that all students at a variety of grade levels should know by the time they complete high school. As such, they provide multiple entry points for the students and, where appropriate, emphasize connections between several areas of mathematics. The discussions point to key teaching strategies that foster the development of reasoning and sense making. The strategies should be viewed as general and not tied to the particular context or task.

Most teachers and teacher educators would probably nod in agreement that reasoning and sense making are important to consider in the mathematical experiences of their students. However, the purpose of *Focus in High School Mathematics: Reasoning and Sense Making* and the Focus in High School Mathematics series is to highlight these as major goals of the study of secondary mathematics. Although reasoning and sense making may have been a part of secondary mathematics teaching and learning in the past, they are certainly worthy of being discussed in greater depth, and becoming a primary focus of our secondary mathematics teaching, in classrooms today. Therefore, with this shift in emphasis, it is important for NCTM to provide thoughtful examples of worthwhile tasks that can be pursued at a number of levels.

The Role of Teaching

Often, high school mathematics teaching in the United States and Canada has been characterized by two main classroom activities; teachers share information, such as definitions of new terms and procedures for solving mathematics problems, and then students practice and perhaps discuss results of those procedures. Although these activities are important, such practices can lead to learning that is devoid of reasoning and sense making. By contrast, NCTM strongly supports a view of mathematics teaching and learning that focuses on reasoning, as described in *Mathematics Teaching Today* (NCTM 2007): "Teachers … must shift their perspectives about teaching from that of a process of delivering information to that of a process of facilitating students' sense making about mathematics" (p. 5).

A shift of perspective to one that views reasoning and sense making as primary goals for students' learning of mathematics will lead to a shift in choices made by the classroom teacher. For example, the teacher will choose tasks that allow students to see the need for sense making and provide opportunities for them to demonstrate their reasoning processes. Such tasks should also help students build on their informal knowledge of mathematics and see the logical connections with other areas of mathematics that they have learned. This shift may require changes in the structure of the classroom setting so that students are challenged and encouraged to explore mathematical situations both collaboratively and independently. Students should be expected to make conjectures and develop arguments to support them, connecting earlier knowledge with newly acquired knowledge.

As students are investigating and shaping ideas, they should have opportunities to interact directly and openly with one another and with the teacher. More details about the teacher's and students' roles in the classroom can be found in chapter 1, "Standards for Teaching and Learning," of *Mathematics Teaching Today*, which includes Standards describing characteristics of *worthwhile mathematical tasks* (Standard 3), components of a productive classroom *learning environment* (Standard 4), and suggestions for orchestrating mathematical *discourse* (Standard 5). The Focus in High School Mathematics series provides tasks, examples, and classroom vignettes that illustrate how a teacher might choose tasks and orchestrate classroom discourse to capitalize on student reasoning and promote sense making.

The Role of Technology

In all of the books in the series, technology is integrated into the examples in a strategic manner to enrich opportunities for students' reasoning and sense making. The power of recent technological tools (e.g., computer algebra systems, dynamic geometry software, and dynamic data representation tools) to enhance reasoning and sense making in mathematics is so great that it would be remiss to omit them from these volumes.

Increasingly, technology is an integral part of society and the research that is conducted in the majority of mathematics-related fields. The series supports the philosophy of *Focus in High School Mathematics: Reasoning and Sense Making* that "students can be challenged to take responsibility for deciding which tool might be useful in a given situation when they are allowed to choose from a menu of mathematical tools that includes technology. Students who have regular opportunities to discuss and reflect on how a technological tool is used effectively will be less apt to use technology as a crutch" (p. 14).

All the companion volumes in the Focus in High School Mathematics series provide examples that show students using technology to reduce computational overhead, while also illustrating the use of technology in experimenting with mathematical objects and modeling mathematical structures. However, NCTM has long recognized the special importance of technology in school mathematics, as expressed in the Technology Principle in *Principles and Standards*. Mathematics education of the highest quality must support students in using technology effectively and confidently. Accordingly, the series develops the topic of technology in a separate volume that highlights the power of technology to assist and advance students' efforts to reason about and make sense of mathematics in grades 9–12.

The Format of the Focus in High School Mathematics Series

Focus in High School Mathematics: Reasoning and Sense Making underscores the need to refocus the high school mathematics curriculum on reasoning and sense making. Companion books provide further insights into how these ways of thinking might develop in three major areas of content in high school mathematics:

- *Focus in High School Mathematics: Reasoning and Sense Making in Algebra*
- *Focus in High School Mathematics: Reasoning and Sense Making in Geometry*

- *Focus in High School Mathematics: Reasoning and Sense Making in Statistics and Probability*

The strand on reasoning and sense making with numbers and measurement discussed in *Focus in High School Mathematics: Reasoning and Sense Making* receives primary attention in *Focus in High School Mathematics: Reasoning and Sense Making in Geometry,* but aspects of this strand are also addressed in the other two content books.

Additional volumes in the Focus in High School Mathematics series develop ideas related to important principles of school mathematics identified in *Principles and Standards.* As noted above, a volume highlighting the use of technology to reason about and make sense of mathematics supports the Technology Principle.

However, no consideration of reasoning and sense making in high school mathematics can be complete without devoting attention to equal treatment of students, regardless of talent, background, and personal advantages or disadvantages of many sorts. *All* high school students must have a chance to reason about and make sense of mathematics in significant ways. Thus, the series includes a volume that highlights equitable opportunities for reasoning and sense making, lending support to the Equity Principle set forth in *Principles and Standards.*

Reasoning Habits

To detail what mathematical reasoning and sense making should look like across the high school curriculum, *Focus in High School Mathematics: Reasoning and Sense Making* provides a list of "reasoning habits." The intent is not to present a new list of topics to be added to the high school curriculum: "Approaching the list as a new set of topics to be taught in an already crowded curriculum is not likely to have the desired effect. Instead, attention to reasoning habits needs to be integrated within the curriculum to ensure that students both understand and can use what they are taught" (p. 9). The reasoning habits are described and illustrated in the examples throughout the companion books in the Focus in High School Mathematics series.

Key Elements

Focus in High School Mathematics: Reasoning and Sense Making identifies "key elements" for each of the strands. These key elements are intended to provide "a lens through which to view the potential of high school programs for promoting mathematical reasoning and sense making" (p. 18), and they are also illustrated in the companion books in the series.

Content Expectations

As *Focus in High School Mathematics: Reasoning and Sense Making* suggests, readers wishing for more detailed content recommendations should refer to chapter 7, "Standards for Grades 9–12," in *Principles and Standards for School Mathematics* (NCTM 2000). However, for the readers' convenience, the appendix of each companion volume shows the relevant Principles or Standards for students in grades 9–12.

Introduction
to *Focus in High School Mathematics:*
Reasoning and Sense Making in Geometry

As the relevance of geometry to our daily lives becomes more apparent, many school systems are requiring, or at least strongly suggesting, that all students take a course in this branch of mathematics. Geometry deals with the shapes of the world in which we live. Geometry courses should not only focus on these (as well as their functions, properties, and measurements), but should also introduce students to the world of thinking and logic, including spatial reasoning, which seems to be so foreign to so many of our young people.

Geometry courses in high school typically range from informal courses that deemphasize proof in favor of exercises that reinforce algebra concepts to honors courses that expect students to master constructions and formal proof writing. In many cases, students who successfully complete any course in this range come out of the experience with a vague notion that geometry is just another course where they must memorize definitions, rules, and properties so that they can apply them in solving a problem similar to those worked in class. For some, geometry means mastering proof writing, or at least learning how to follow the teacher's model for writing appropriate statements and reasons in the correct order to show the validity of some geometric property. But formal deductive proof writing should not be the main goal of a geometry course.

Formal proof has a place in the geometry curriculum. Formal proofs can help students to develop ways of thinking about, questioning, and justifying mathematical situations. They are useful as students consolidate and verify their reasoning for a particular problem situation. But thinking, questioning, and justifying should occur whenever students encounter a situation that is new to them, both within and outside of the school setting, and not only when a proof is required in geometry class. We rarely need to line up statements and reasons in our daily life, but we frequently need to provide a rationale, if only for ourselves, as to why we do or say particular things. Reasoning and sense making should be regular parts of the geometry curriculum, with or without formal proof writing. All students, not just those who seem to have an "intuitive" comprehension of mathematical ideas, must be able to reason and make appropriate decisions. Geometry provides an environment that can allow and encourage students to "practice" the process of reasoning and sense making, for the benefit of us all.

As identified in *Focus in High School Mathematics: Reasoning and Sense Making* (NCTM 2009, pp. 55–56), four key elements of reasoning and sense making in geometry include the following:

- *Conjecturing about geometric objects*. Analyzing configurations and reasoning inductively about relationships to formulate conjectures.

- *Construction and evaluation of geometric arguments*. Developing and evaluating deductive arguments (both formal and informal) about figures and their properties that help make sense of geometric situations.

- *Multiple geometric approaches*. Analyzing mathematical situations by using transformations, synthetic approaches, and coordinate systems.

- *Geometric connections and modeling*. Using geometric ideas, including spatial visualization, in other areas of mathematics, other disciplines, and in real-world situations.

Why are these key elements of geometric reasoning so valuable? What should be the role of reasoning in high school geometry? The answer to the first question is easy. Inherently, as human beings, we want to make sense of the world around us. We do this by connecting new ideas to existing ones, or by readjusting our thinking so that new ideas fit with what we already know or believe. In the geometry classroom, this kind of sense making involves geometric reasoning—purposefully making inferences, associations, or deductions about and among geometric concepts, objects, and structures (Thompson 1996).

In geometry, at all levels of schooling, reasoning must involve active exploration of shapes so that students can investigate attributes of a shape, common properties of a family of shapes, or a variety of ways to model shapes (Clements and Battista 1992). Such explorations provide a basis for rich geometric reasoning. Dynamic geometry environments (DGEs), supported by a range of computer programs, allow for and encourage exploration. Judicious use of DGEs can significantly enrich geometric reasoning in a secondary school curriculum. However, research has shown that many students do not feel the need to ask why or justify (prove) the validity of their conjectures once they "see" the truth of geometric relationships (Battista 2007; Hershkowitz et al. 2002). Thus, student explorations provide only a first step in sense making. Geometric reasoning goes beyond conjecturing; it involves finding inductive and deductive ways to make sense of relationships among and within geometric objects. The tasks in which students engage, the classroom environment and tools available, and the nature of the classroom discourse all play parts in developing students' sense making and reasoning in geometry (NCTM 2007).

The Themes and Flow of This Book

The four key elements that *Focus in High School Mathematics: Reasoning and Sense Making* (NCTM 2009) identifies for reasoning and sense making in geometry are a major focus of this book and should be a major focus in any geometry course. However, they also are critical across the high school mathematics curriculum. Developing conjectures and constructing or evaluating mathematical arguments are valuable experiences for mathematics students at any grade level and in any content area. Mathematical modeling and using multiple approaches or representations are common themes promoted by the National Council of Teachers of Mathematics for all students (NCTM 2000). In this book, we explore these key elements in relation to the teaching and learning of core geometry topics in the high school curriculum.

The concepts of congruence and similarity play a major role in the geometry curriculum. An understanding of these relationships can lead to more sophisticated investigations in geometry. Chapter 1 provides two tasks that involve congruence and similarity along with other related topics, such as transformations, measurement, proportional reasoning, and algebraic connections.

Analyzing and exploring characteristics and properties of shapes and space are very traditional components of the high school geometry curriculum. The study of geometric objects, whether they are two-dimensional or three-dimensional, is relevant to students, connecting them to the world in which they live. When students are able to recognize relationships among objects, they are more likely to develop deeper knowledge of them and become able to solve problems or analyze new situations involving these shapes and their properties and relationships in space. Reasoning and sense making with two- and three-dimensional objects and their properties are the focus of chapters 2 and 3.

Geometric modeling is the focus of chapter 4. A modeling example shows how an authentic problem situation can be developed to introduce a range of students to the mathematics modeling process and promote an investigation of related geometric properties. A real-world problem is explored through a series of tasks that require geometric reasoning about shapes, dynamic relationships among shapes, and reasoning about concepts from other mathematical content areas to aid in solving the problem.

Each chapter provides tasks, classroom scenarios, and examples of possible lines of student reasoning. The intent is to describe possibilities for focusing or refocusing on reasoning and sense making in the geometry classroom. The tasks have been chosen as examples that have the potential to promote geometric reasoning at a variety of levels, for students with a range of abilities. The classroom vignettes are based on actual teacher-student interactions that have occurred in implementing the tasks in geometry classrooms. Thus, teachers who are reading the book may get a sense of what is possible. These vignettes and accompanying notes to teachers also provide insight into choices that other teachers have made during classroom activities about whether to highlight students' reasoning and how to promote or enhance reasoning among students. As a result, the book can serve as a starting point for some and a resource for all about reasoning and sense making in geometry.

Chapter 1

Reasoning about Congruence and Similarity

Congruence and similarity are central relational concepts in the study of geometry. An understanding of these relationships provides students with tools to investigate and analyze other relationships among, and properties of, shapes (e.g., transformations and how they function). These geometric relationships help to connect many concepts within geometry and to link geometry itself to other areas of mathematics and to problems in the world around us. For instance, the concept of similarity is closely tied to proportional reasoning, scale factors, growth and decay, and indirect measurement. These and other connections make the study of congruence and similarity central to the geometry curriculum.

The two tasks presented in this chapter center on congruence and similarity in that an understanding of these concepts is crucial for making sense of the tasks and developing solid reasoning about the solutions to the associated problems. Other geometry concepts involved in the first task, Rotating Square, include ideas related to polygon angle sums, rotations, and area. The chapter's second task, Field of Vision, involves connections to measurement, data analysis, and linear functions.

Although the teachers and classroom episodes that the chapter presents are fictitious, these activities have been used in many classrooms, with a wide range of students. The episodes reflect actual student reasoning and teacher direction or guidance, although they have been slightly idealized and made smooth for easy reading.

Rotating Square

The Rotating Square problem presents two congruent squares in the configuration shown in figure 1.1.

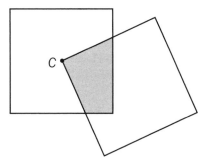

Fig. 1.1. Rotating Square problem

Rotating Square

Two congruent squares (n units by n units) overlap as shown in figure 1.1. Vertex C of one square is at the center of the other square. If the square with vertex C is allowed to rotate about the center, C, of the other square, what is the largest possible value of the overlapping shaded area?

In the classroom

Mr. Lee asks the students in his geometry class (a mix of students in grades 9, 10, and 11) to investigate the situation in the Rotating Square task and develop a conjecture. He also asks them to think about how they might justify or explain their conjectures to classmates. Mr. Lee assigns his students to groups of three to work on the task.

Students in group 1 immediately begin drawing other possible positions for the rotating square and notice that at one point the area of overlap will be a square that has an area that is exactly $\frac{1}{4}$ of the area of the original square. A group member guesses that this will be the maximum area because in a previous activity where the students investigated the greatest area in rectangles with a set perimeter, they discovered that the square had the largest area (see fig. 1.2 for an example).

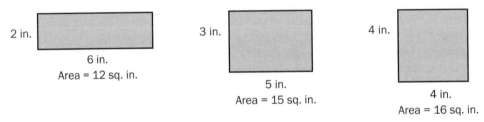

2 in.   6 in.   Area = 12 sq. in.

3 in.   5 in.   Area = 15 sq. in.

4 in.   4 in.   Area = 16 sq. in.

Fig. 1.2. Rectangles with perimeters of 16 inches

In group 2, one student is having difficulty visualizing how the shape of the shaded region will change as the top square rotates. Another student cuts out two congruent squares from grid paper and uses the tip of her pen to hold a vertex of the rotating square at the center of the other square. She shows the others in her group how the overlapping shape changes, using her physical model as an aid. A group member wonders if the group could find a way to count the grid squares on the paper to find the area of overlap between the two squares. Mr. Lee notices the physical model constructed by group 2 and asks the members to share it with the rest of the class. He thinks others might find it helpful to build such a model.

After another ten minutes of exploration and discussion, Mr. Lee asks all the groups to share their ideas and initial conjectures with the whole class:

Kelsey: [*Speaking from group 1*] We tried several spots for the rotating square. We saw that it [*indicates the shaded region in fig. 1.3*] can become a square, or a triangle, or just some general quadrilateral. When it's a square and a triangle, we found that the shaded area is $\frac{1}{4}$ of the big square. We think this is as big as it can get.

Tolu: [*Speaking from group 3*] We saw that, too, but we weren't sure how to find the area of the other four-sided figures. Maybe they're bigger than the square and triangle.

Mr. Lee: OK. Did anyone find a way to calculate the area of the more general quadrilateral? Can you help out the students in group 3? Perhaps someone else has an idea about this?

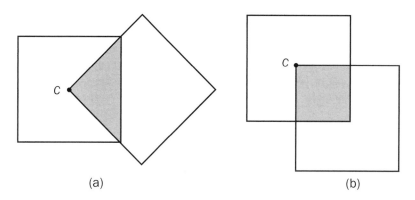

Fig. 1.3. Two regions formed by overlapping squares: (a) triangle and (b) square.

Nicole: [*Speaking from group 4*] I have a different but related question. How do you know that the triangle shape is $1/4$ of the square? How do you know that when you rotate the top square, it'll make a triangle that crosses exactly at the corners of the bottom square?

Kelsey: Oh, I can answer that because I had that question, too. So I thought about it, and [*holding up her drawing to show others in the class; see fig. 1.4*] I noticed that if you connect the center of the bottom square [*indicating the stationary square*] to two of the corners of that square [*showing how her group used segments*], you get a right angle at the center. And the top square [*indicating the rotating square*] has a right angle, so it must fit exactly in that space if you rotate it just the right amount. So its sides must go from the center to the corners of the bottom square.

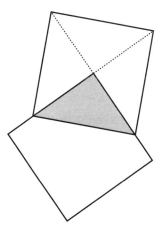

Fig. 1.4. Kelsey's drawing showing the triangle as $1/4$ of the square

Mr. Lee: Good observation. [*To Nicole:*] Do you see that? [*To the class:*] Does everyone see that? Remember, we've seen that the diagonals of a square are perpendicular. In this example, Kelsey is using that information to justify that the rotating square can be placed so that it lies exactly within two pieces of the diagonals, and the shaded region will be a triangle that is $1/4$ of the bottom square. OK, so what about when the shaded region is a more general quadrilateral? Did anyone find a way to get the area?

Abram:	[*Speaking from group 2*] I think we've got something. We made these two squares out of graph paper [*holds up the physical model created earlier*]. And we held this corner in the center of the other square and turned the top square around. Then we tried to estimate the area of where they overlap by counting all these little squares on the paper. We could be wrong, but we think that the area is the same no matter how you rotate the top square. Our numbers didn't match exactly, but they were almost the same all the time, and we were just estimating.
Nathan:	[*Speaking from group 5*] We did it a different way, but we came up with that, too—that the shaded area is always $\frac{1}{4}$ of the big square on the bottom. We saw that when you rotate the top square, the amount that you move it in one direction is the amount that you lose in the other direction. Um, I don't know if that makes sense… It's hard to explain.
Mr. Lee:	Can you show us? Up here [*motions for Nathan to come to the front of the classroom*].
Nathan:	Um, I can try. [*Goes to the front of the class and uses the sketch that Mr. Lee drew on the board initially. He draws a few extra segments, as shown in fig. 1.5. Pointing to the perpendicular segments indicated by dashed lines in the figure, he shares his observation.*] We drew these lines because we wanted to find the area [*indicating the shaded region in the figure*] by using triangles, and we needed the heights of the two triangles. Then I noticed that these little triangles, one inside and one outside of the shaded region, are the same. So, we think that if you start with the square [*indicating the overlapping region as a square*] and then rotate it to the right [*motioning counterclockwise*], then the amount you rotate… the amount you gain is the same as what you lose down here [*points to the lower triangular region that is outside the shaded region and the analogous upper triangular region that is inside the shaded area*]. So the area is always the same.

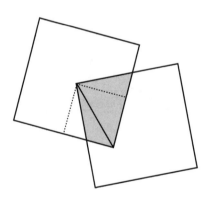

Fig. 1.5. Nathan's drawing

Several students:	Oh, I see it now. I agree with Nathan. I think that area will always be the same.
Mr. Lee:	So, who thinks we can write a conjecture about this situation? Are we ready to put it together? Tolu?
Tolu:	Yeah. When two congruent squares overlap, like in the drawing, then the amount they overlap is always the same, even if the top square rotates around the center. Oh, and the area is $\frac{1}{4}$ of the bottom square.

Discussion of the students' work on the Rotating Square problem

As shown in the classroom investigation, students can use a variety of methods to approach the Rotating Square problem. Students who need help in visualizing the situation can quickly build a physical model with paper or cardboard. Other students may be able to work with paper and pencil without manipulating a model, while yet others may be able to visualize the situation in their heads, without the help of a drawing. Another useful tool is an interactive geometry software package. A dynamic computer model is ideal because it allows students to construct a drawing like that shown in figure 1.1, measure the desired area, and quickly find the "answer" to the question of maximum area. Students who are skillful with interactive geometry programs may quickly construct the two squares, while just constructing one square could be challenging for other students. Constructing a congruent square with a vertex at the center of the first square so that the second square rotates may also challenge some students, but doing so can be a great way to support or introduce congruence of geometric figures.

When students create models in any medium, questions arise that they will need to answer for themselves or with the help of peers or a teacher. For instance, students will need to understand how to find the "center" of the stationary square. They might ask questions like, "How do you define the center of a polygon? Is it the same for all polygons? Only regular polygons?" or, "How do you locate the center of a regular polygon? Is the process for finding the center different for different regular polygons? How can you check that you have found the center?" Students should also be prompted to consider other questions, such as whether the shaded region will always be a quadrilateral, and whether it can ever be a parallelogram.

Another interesting question that students might answer by exploring and considering properties of squares is that raised by Nicole in the classroom scenario: Could the shaded region possibly form a triangle that does not intersect at the vertices of the stationary square? A formal proof of the proposition that sides of the triangular overlapping region intersect adjacent vertices of the stationary square may not be desired. However, the informal argument given by Kelsey and the teacher helps all students make connections to previous concepts and justify an assumption that many of the students might have made. In this particular case, Kelsey drew in the diagonals of the stationary square and found or remembered that these diagonals will form a right angle. Her *analysis* of the situation, using not only given information but also *hidden structure*, allowed her to make a useful *connection* to answer a question and *justify its validity*. These are all valuable reasoning habits that teachers should foster in the classroom on a regular basis. Mr. Lee then supported Kelsey's response by clarifying and stating previously learned relationships (e.g., diagonals of a square are perpendicular).

Although the students answered the question of how to find the area of the general quadrilateral indirectly, by realizing that it is the same as the area of the small square, this question can also be something to explore in greater depth with students. For instance, if Nathan had not made the discovery about congruent triangles within the figure, the teacher could have pushed the students further by helping them find the area of the more general quadrilateral. Finding the area of more general shapes is often possible by breaking them down into more familiar shapes whose areas are easier to calculate. Nathan and his fellow group members attempted to divide the quadrilateral into two triangles. Another approach would be to divide the quadrilateral into a right triangle and a trapezoid by dropping a perpendicular from the center of the stationary square to one of its sides (see fig. 1.6a). Yet another approach would be to create a rectangle by adding right triangles to the shaded quadrilateral, finding the area of the rectangle, and subtracting the area of the added triangles (see fig. 1.6b). Students may also recognize that these two smaller, added triangles are congruent right triangles. Sharing a variety of approaches will allow students to pick an approach that seems most reasonable for them.

The questions and possible discussion points suggested in the preceding paragraphs are just a few of the questions that students need to be thinking about themselves. Answers to these questions will help students make sense of the situation under investigation. Without a solid understanding of what is

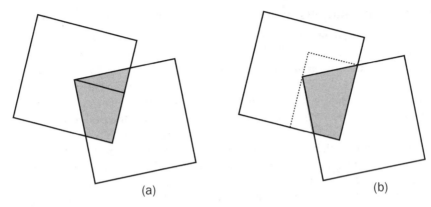

(a)   (b)

Fig. 1.6. Two other approaches for finding the area of the shaded region

happening and why, students will not be able to develop the deeper reasoning that they need to justify the problem solution. Table 1.1 highlights some of the key elements of geometry and reasoning habits (NCTM 2009, pp. 9–10, 55) illustrated by this problem and the students' initial work on it.

Table 1.1
Key Elements and Reasoning Habits Illustrated in the Rotating Square Problem

---

Key Elements of Reasoning and Sense Making with Geometry

Conjecturing about geometric objects
  Reasoning inductively and deductively by using a variety of representations
Construction and evaluation of geometric arguments
  Developing arguments (informal) to justify a conjecture
Multiple geometric approaches
  Analyzing a situation by using transformational and synthetic approaches

---

Reasoning Habits

Analyzing a problem
  Looking for hidden structure by drawing auxiliary lines
  Seeking patterns and relationships
  • Systematically examining cases
  • Considering special cases
  Making preliminary conjectures
Reflecting on a solution
  Justifying or validating a solution through informal proof

The next section shows the students in Mr. Lee's class moving beyond the conjecture and working together to formulate a formal justification (or proof) of it. Although a formal proof is not always the goal of a lesson and may not be an appropriate expectation for all students, the glimpse provided here of students formulating a proof of the Rotating Square conjecture suggests what is possible when reasoning and sense making are fostered in the mathematics classroom.

Proving the conjecture about the Rotating Square problem

Mr. Lee asks his students to take the next step in the investigation by justifying the conjecture that they formulated in exploring the situation illustrated in figure 1.7:

Proving the Conjecture

Justify the conjecture from the Rotating Square exploration—namely, that the area of the shaded region is always $\frac{1}{4}$ of the area of the non-rotating square with center C and does not depend on the shape of the shaded region.

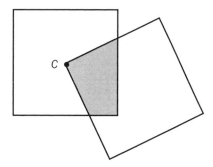

Fig. 1.7. Rotating Square conjecture: The area of the shaded region is always $\frac{1}{4}$ of the area of the non-rotating square with center C.

In the classroom

Mr. Lee writes the conjecture on the board and sends the students back to their groups to develop a way to justify the conjecture. He asks the students to use what they have seen and heard from classmates to find a way to convince others—perhaps someone who is not in the classroom—that the conjecture is true.

The students in group 1 return to their original work and decide to show a variety of cases—that is, various positions for the rotating square. They plan to calculate the area of the overlapping region for each case and show that it is always $\frac{1}{4}$ of the area of the stationary square. The students in group 3 follow this same line of thinking. They are discussing ways of finding the area of the more general shaded region. One student wonders if showing that the area of this region is $\frac{1}{4}$ of the area of the stationary square in three different cases would be enough—that is, when the overlapping region is (a) a square, (b) a triangle, and (c) a more general quadrilateral, as shown in figure 1.8.

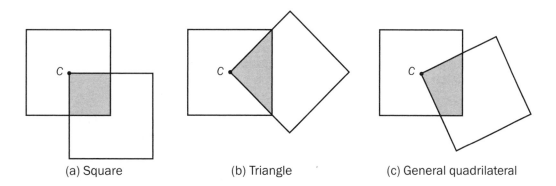

(a) Square   (b) Triangle   (c) General quadrilateral

Fig. 1.8. Three possible cases for the shaded region

A student in group 2 wonders if the areas that the group estimated by using the grid paper would be convincing enough. The group members are not sure about this method. They have found it difficult to count all the tiny grid squares and then make estimates about the squares that are not fully enclosed in the shaded region (see fig. 1.9). The group decides to explore other ways to justify the conjecture.

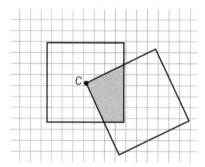

Fig. 1.9. Using grid paper to find area

Groups 4 and 5 are working together, sharing ideas related to Nathan's discovery that the amount of area that the shaded region loses is equal to what it gains as the rotating square moves from one position to another. Nathan redraws his diagram, leaving out a few unneeded line segments and labeling important intersection points, as shown in figure 1.10, and the students have the following discussion:

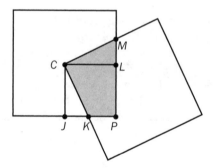

Fig. 1.10. Nathan's drawing with labels

Nicole: OK, I see now that when we have the square shape and the triangle shape, the overlapping area is $\frac{1}{4}$ of the bottom square's area. But I still don't get what you were saying, Nathan, about losing some area and gaining some area when you rotate it.

Xhen: I can show you, I think. Look at the square here, with vertices *C, J, P,* and *L*. So we have square *CJPL*, and that's the overlapping area. But when you move the top big square by this amount [*points to ∠JCK formed by rotating the top square counterclockwise*], it moves the same amount on the other side [*points to ∠LCM*]. So we have triangle *JCK* and triangle *LCM,* and they have the same areas. We just have to convince someone else that we're right, like Mr. Lee said.

Nathan: Well, those two small triangles have the same height. Because there are right angles at *J* and *L*, and segment *CJ* has the same length as segment *CL*. They are both half the length of a side of the square. But how can we convince someone that the bases of the triangles are the same? Why is *JK* equal to *LM*?

Will: They just are. It's like what Xhen said. When you move the top square this amount, it moves the same amount along the other side.

Nathan: But you know what Mr. Lee is going to say… "Why?" "How do you know?" How do we know for sure that segments JK and LM are the same length? Have we seen this before somewhere?

Nicole: Hmm… I see the two triangles, JKC and LMC. Can we show that these triangles are congruent by using different parts? Nathan said segments CJ and CL are congruent. And both triangles have right angles at J and L. Now, what's the third piece to show the triangles are congruent, besides the segments JK and LM? What about another pair of angles, like angles JCK and LCM?

Xhen: Yes, I think you're right. Angle JCK and angle KCL make a right angle. And so do angle LCM and angle KCL. So if we subtract the measure of angle KCL from both right angles, we get these two angles (points to $\angle JCK$ and $\angle LCM$) with the same measure.

Olivia: I think we can put it all together now. If we show that these two small triangles are congruent, using angle-side-angle, then we've shown that the area of the shaded piece is always the same. It's like Nathan said, no matter how much you turn the top square, what you lose in triangle JKC, you will gain in triangle LMC.

Discussion of the students' finding of the proof

The follow-up activity required students to think more deeply about the conjecture related to the Rotating Square investigation. All the groups had ideas for attempting a justification of the proof. The students based these ideas on their previous work in developing the conjecture and on the class sharing and discussion that followed their initial investigations. Although the justification developed by groups 4 and 5 in the vignette above follows a particular method—a synthetic approach to a formal proof—other groups appear to be moving in slightly different directions. Two approaches that are not explicitly mentioned by students or the teacher in the scenario above are an analytic (or coordinate) approach and a transformational approach.

The analytic approach to proof makes use of the xy-coordinate plane. Such an approach uses algebra and arithmetic operations and helps to make connections with students' prior work in these areas. A transformational proof would use the fact that the top square is rotating about a given center of rotation. Students can then argue that the segments CJ and CL in Nathan's diagram are rotated an equal amount about the center C, so the measure of angle JCK is equal to the measure of angle LCM. This is an alternate way of showing what Olivia accomplished by using subtraction of angles or complementary angles. A second transformational argument could follow from four successive 90° rotations of the shaded region, demonstrating that the shaded region is $\frac{1}{4}$ of the entire bottom square.

The focus of this task is to help students develop and evaluate their own geometric arguments. Thus, as students work together, they should question their own ideas as well as those of others, rather than accepting what appears to be true, without proof. This process is evident in Nathan's questioning of himself and his group members about why the segments JK and LM can be said to be the same. Echoing a familiar remark from the teacher—"How do you know?"—Nathan encouraged his group members to continue searching for a strong justification of the conjecture. That simple question, "How do you know?" is an important one for helping students use key elements of reasoning and sense making and develop valuable reasoning habits, as summarized in table 1.2 (NCTM 2009, pp. 9–10, 55).

Table 1.2
Key Elements and Reasoning Habits Illustrated in Proving the Conjecture

Key Elements of Reasoning and Sense Making with Geometry

Construction and evaluation of geometric arguments
Considering the role of empirical evidence
Developing a formal deductive argument to determine mathematical certainty

Multiple geometric approaches
Analyzing a situation using transformational and synthetic approaches

Reasoning Habits

Analyzing a problem
Seeking patterns and relationships

Implementing a strategy
Organizing ideas for a solution
Making logical deductions based on current progress

Reflecting on a solution
Considering the reasonableness of a solution
Justifying a solution through formal proof

Field of Vision — Making Sense of Similarity

In the investigation that follows, students explore the Field of Vision problem (adapted from the Viewing Tube problem [Cooney et al. 1996, p. 46]):

> **Field of Vision**
>
> At one time or another, most of us have used a cardboard or plastic tube as a telescope. Although the tube does not actually enlarge what we see, it does help us focus on a narrow field of vision. In this problem, you will explore relationships among the variables associated with such a viewing tube and the field of vision afforded by the tube.

The students approach this problem through three tasks that allow them to work in various mathematical domains.

Task 1

In the first task associated with the Field of Vision problem, hands-on work with viewing tubes engages students in data collection and measurement and gives them a brief look at patterns and algebra:

> **Task 1:** Using the tube given to you, collect data to determine a relationship between the distance that a viewer's eye is from a vertical wall and the viewer's field of vision on that wall.

It is helpful for all students to use tubes of the same length and diameter in task 1. The students in different groups can then compare or compile data to make conjectures. Before the students collect their data, teachers might define the term *field of vision* as the area of the circle on the vertical wall that is visible through the tube. Or, to simplify the mathematics, they might choose to define it as the diameter (or radius) of the circle visible through the tube. A third option would be to let each group of students decide how to define and measure *field of vision*.

In the classroom

Ms. Van Ledtje distributes paper-towel tubes, masking tape, and meter sticks to each group of three or four students in her geometry class. With these materials, as well as paper and pencils to record the data, each group finds an open space on the classroom wall. Ms. Van Ledtje suggests using the tape to represent the diameter of the field of vision. Most groups put a long strip of tape on the wall, either parallel or perpendicular to the floor.

Data collection strategies vary among the groups, but most incrementally increase the viewer's *distance from the wall* (independent variable) and use various techniques for measuring the viewer's *field of vision* on the wall (dependent variable). A discussion about measurement techniques arises between two neighboring groups. One student, Steven, notices that the students in the next group are measuring the distance from the viewer's eye to the wall, but his group is measuring the distance from the viewer's toes to the wall. Steven asks the members of the other group why they chose to record these measurements. Ms. Van Ledtje overhears the conversation and stops all the groups to share Steven's observation and question and to reach a consensus on how to collect accurate data for this task. It turns out that the groups are measuring four different lengths (see fig. 1.11):

(*a*) Distance from viewer's eye to wall

(*b*) Distance from end of tube (away from eye) to wall

(*c*) Distance from viewer's toes to wall

(*d*) Distance from viewer's heels to wall

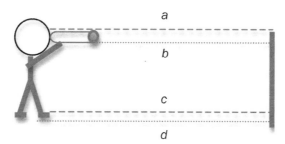

Fig. 1.11. Which distance is most useful to measure?

After discussing the matter briefly and rereading the task, all the students agree that the distance should be measured from the viewer's eye to the wall. This conversation also uncovers two assumptions—that the viewer's body is parallel to the wall and that the viewer's line of sight is perpendicular to the wall (i.e., the head is not tilted down or up).

Within another five minutes, most groups have completed their data collection and are beginning to look for a relationship between the viewer's *distance from the wall* and the viewer's *field of vision*. Leo, Khahn, and Melissa decide to plot the data points on a coordinate grid, with distance from the wall represented on the *x*-axis and diameter of field of vision represented on the *y*-axis. They discuss their resulting plot:

Melissa: Well, it's not exactly a line [*gesturing toward the scattered points on the graph paper in front of her*], but I'd say that we have a linear relationship. We just aren't very accurate with measuring. Maybe we should try again and check our measurements. I bet it would be more like a line if we had better measurements.

Khahn: What's the equation of the line? Let's just write it and check with Ms. V. I don't want to measure everything over again.

Leo:	Wait a minute—I have two questions. First, this is geometry class, so why are we graphing data points and looking for an equation of a line? That's like algebra. Second, if we're supposed to find a relationship between distance and field of vision, aren't we supposed to use πr^2, and not just the diameter of the field of vision? That would bring in some geometry.
Melissa:	I see what you mean. But that would give us a different relationship, something like a parabola with the r squared. I think the line is good enough.
Khahn:	Yeah, I agree. We could use the points we've graphed to find the real "field of vision" by doing the πr^2 thing. The line gives us the information we need to do that.

[*Khahn continues to search for a linear equation to fit the data points. He lightly sketches a line that passes close to the origin and connects about three of the eight data points. The other five data points are scattered close to the line, four above the line and one below it. He uses two of the data points on his line to calculate slope.*]

Khahn:	I think the line is like, $y = 0.2x$ plus something small. Let's see what Steven's group has.
Melissa:	Yes, that makes sense. Every time I stepped back 20 centimeters, I could see about 3 or 4 more centimeters of the tape on the wall. That's like the slope of the line, $^4/_{20}$, or 0.2. Your equation must be right.

Discussion of students' work on task 1

At the beginning of the investigation, the teacher chose to leave data collection methods up to the students. Although students quickly found methods for gathering data, they soon became aware of some variation in their methods. The teacher decided to discuss measurement techniques with them before they finished the data collection. This helped the students refocus on the statement of the task and make good choices about the appropriateness and *reasonableness of their measurements*. Some students may require more guidance through the data collection portion of the task. The teacher's suggestion about using the tape to represent the diameter of the field of vision is a good start, but students may also need direction about how to vary the independent variable—the distance to the wall—and how to measure for the dependent variable—the diameter or area of the field of vision. Ms. Van Ledtje's more open-ended approach required the students to make sense of the physical situation and to find methods for mathematizing it.

The students in Melissa's group quickly turned to a graphical model of the data to *seek relationships*. This model helped Melissa and Khahn attach mathematical meaning to the physical situation. Melissa took this process a step further and attempted to make sense of Khahn's algebraic representation by connecting it to the data and to the physical situation that she experienced. In this way, she was able to *check the reasonableness* of Khahn's solution by connecting slope to the data collection variables, change in distance and change in diameter of field of vision (*meaningful use of symbols*). Table 1.3 shows the key elements and reasoning habits (NCTM 2009, pp. 9–10, 21, 31) illustrated by this approach to task 1.

Table 1.3
Key Elements and Reasoning Habits Illustrated in Task 1

Key Elements of Reasoning and Sense Making with Measurement and Algebra

    Reasonableness of measurements

        Judging whether a measurement has an appropriate order of magnitude

Table 1.3—*Continued*

Approximations and error
Realizing that all real-world measurements are approximations
Meaningful use of symbols
Choosing variables and constructing expressions in context
Connecting algebra with geometry
Representing geometric situations algebraically

Reasoning Habits

Analyzing a problem
Seeking patterns and relationships
Making preliminary deductions and conjectures
Implementing a strategy
Making purposeful use of procedures
Organizing a solution through data displays and calculations
Reflecting on a solution
Considering the reasonableness of a solution

At this point, all the students in Ms. Van Ledtje's geometry class agree that the relationship between D_W (distance from the wall) and F_W (*diameter* of viewer's field of vision on the wall) is linear with a slope of about 0.2. They also agree that the slope represents the ratio of change in their field of vision to change in distance from the wall. But the students do not agree on whether the line representing the data passes through the origin. Some say the y-intercept is equivalent to the diameter of the tube because the viewer will always be able to see at least that much of the wall. Others disagree, noting that 0 centimeters from the wall ($D_W = 0$) means not using a tube and thus having no field of vision ($F_W = 0$).

Task 2

Ms. Van Ledtje decides to move on and allow students more time to think about their different perspectives. On the front board, she writes task 2, which brings algebra to the forefront:

> **Task 2:** Study your original viewing tube and other tubes of varying lengths and diameters. Identify all variables that may influence the relationship that you determined in task 1. Find a general solution so that the field of vision can be calculated for a tube of any size.

In the classroom

Ms. Van Ledtje asks her students to begin by considering all variables that influenced the data that they collected. To help the students, she displays other tubes that are shorter, longer, wider, or narrower. Students begin shouting out variables, such as distance from the wall, diameter of the tube, and length of the tube.

Ms. Van Ledtje: OK, you all have some data about the original tube. And now you have several more tubes and a list of potential variables. What suggestions do you have for finding a more general solution?

Steven: Each group can take a new tube and collect new data. Then we can pool the data and look for a general pattern across the tubes.

Julisa: The pattern will have more than two variables. We've listed four! What kind of equation has four variables? It can't be linear.

Steven:	I think it can still be a line, but some of the variables can be used for *y*-intercept or slope, like our first equation did [*points to the equation that his group shared: y = 0.1x + 3.5, where the y-intercept of 3.5 represented the diameter of the tube in centimeters*].
Ms. Van Ledtje:	[*Addressing the entire class*] What do you think about Steven's suggestion that we collect more data and see what it tells us?
Several students:	Yeah. Good idea.
Steven:	But we need to know which tube the data goes with. Let's number the tubes.
Khahn:	No, let's record the diameter and length of the tube. We said those were other variables to consider. So maybe those measurements have something to do with the new data we'll collect.
Ms. Van Ledtje:	Good observation. Be sure to record which tube you are using for your new round of data collection. Record the tube's length and diameter.

Note that Ms. Van Ledtje introduces task 2 but does not immediately let students start on it. She first asks them to consider the problem and brainstorm strategies for solving it. This gives all students time to analyze the problem and identify relevant information.

Ms. Van Ledtje randomly distributes the new tubes and gives the groups ten minutes to record five or more data points. As the students begin collecting more data, Khahn makes the following observation to Melissa and Leo: "I think we can do it without collecting any more data. Like Steven said, the other variables are just the other relationships in the equation. Let's get the original tube back and measure its length and diameter. Maybe the ratio of these is the *y*-intercept."

Melissa and Leo like Khahn's suggestion, but they know that they must also collect new data to share with the class. They collect data with the new tube while Khahn tests his idea. Students from all groups then share the new data with the class.

Task 3

After everyone records the new data in their notebooks, Ms. Van Ledjte writes task 3 on the board:

> **Task 3:** Use the new and old data to find a general relationship among the four variables identified as follows: length of tube (L_T), diameter of tube (D_T), distance from wall (L_W), and diameter of field of vision on wall (D_W). Write a careful justification for this relationship. *Hint:* Try modeling the physical situation with geometry.

This task, which will be the students' homework, relies on geometric modeling, use of similar triangles, and geometric reasoning.

In the classroom

At the start of the next class, the students get back into their groups from the previous day to share their solutions and justifications. Ms. Van Ledtje asks each group to come to an agreement on the best solution and justification and record their solutions on overhead transparencies to share with the class. The groups ultimately present three different lines of reasoning. One strategy relies on algebra, and the other two rely on geometry.

Algebraic strategy

Julisa's group (and one other) used their graphing calculators to find a line of best fit for each set of data related to the different tubes. They then looked for patterns among the linear equations and a relationship to the size of the tubes. They concluded that the diameter of the tube is proportional to the slope of the line. That is, a larger diameter will give a greater value for the slope of the associated line. They justify this conclusion to their classmates by explaining that a wider opening gives a larger diameter for the field of vision, so the slope must be greater.

The students generally agree about this conclusion. Julisa has written each linear equation on the transparency, arranging all the equations in a chart according to the size of the tube's diameter. It is easy to see that the conclusion is true for the data. However, Nick and others notice that two tubes that have the same diameter are associated with linear equations that have different slopes. The shorter tube has a greater slope. Julisa quickly jumps in with a justification.

Julisa:	Yeah, we saw that, too. If the tube is shorter, you don't get as much of your vision blocked, so it's like having a wider opening.
Nick:	So, you're saying that the slope depends on both the size of the diameter and the length of the tube. Because that's what we said. We have a different way of showing it.
Julisa:	Yeah, I guess you can say that. I'm not sure how to write it. The shorter tube gives a bigger field of vision, so that's inversely proportional, right, Ms. V? I'm not sure how to account for that in our table. Maybe we can look at the ratio of diameter and length of the tube?
Nick:	That's what we tried to show, but the numbers aren't quite right.
Ms. Van Ledjte:	That sounds like something to check out. How about if you hold on to that idea while we look at the other solutions?

Geometric strategy 1

Leo's group shares a different approach. The members of the group investigated Khahn's idea from the previous day—the idea that the y-intercept is the ratio of the tube's diameter to its length. They noticed that this relationship doesn't work, but they realized instead that the ratio of tube diameter to tube length gives an approximate value for the slopes of the lines that they have found. Heeding Ms. Van Ledtje's suggestion about using geometry, Leo has decided to use similar triangles to justify Khahn's conclusion about the slope because similar triangles are proportional, just as the two ratios of variables appear to be proportional ($D_T/L_T = D_W/L_W$). Leo shares a diagram like that in figure 1.12 and explains his reasoning to the class.

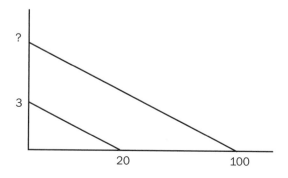

Fig. 1.12. Leo's diagram

Leo:	I put the diameters on the vertical axis and lengths or distances on the horizontal axis. So the 3 means that the diameter of the tube is 3 centimeters, and the 20 is for the length of 20 centimeters. So, to find the field of vision if you are a meter from the wall, that's 100 centimeters… You put 100 on the horizontal axis and find the missing value in the similar triangles.
Ms. Van Ledjte:	Does anyone have questions for Leo?
Julisa:	I do. How did you know that you should create similar triangles?
Leo:	Just like I said earlier, the slope is the ratio of diameter to length of the tube. We showed yesterday that the field of vision and distance from the wall also represent slope of the line, so they will be equal ratios. So, similar triangles.
Julisa:	I get that, but why did you draw the triangles that way?
Leo:	That's the only way I could think of to compare diameter to length.
Ms. Van Ledjte:	Leo, or anyone else who thought of it this way, can you explain to others how you know the triangles are similar?
Leo:	I drew them that way. I drew them so they would be similar, because the equations show the proportions I talked about.
Ms. Van Ledjte:	Hmm… Is there something else about the data collection or about how you analyzed the data yesterday that relates to similar triangles?
Melissa:	That part confused me, too. I'm in Leo's group so I wanted to make sense of what he was saying, but I wasn't sure if we could just draw these similar triangles because we knew there should be similar triangles. I wanted it to make more sense… Like we draw similar triangles because that's what we're doing. But my solution didn't make sense either. Leo's is better.

Geometric strategy 2

Matt and Anna put their transparency on the overhead projector (see fig. 1.13). It also has similar triangles, but the drawing is different from the one that Leo showed. Anna describes how the diagram models the data collection situation.

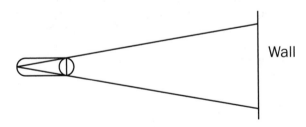

Fig. 1.13. Matt and Anna's drawing of the similar triangles in the data collection situation

Anna:	So here is the tube [*points to the drawing*], and that makes the small similar triangle. It goes from your eye to the other end of the tube. The big triangle goes from your eye all the way to the wall.
Matt:	And we know the triangles are similar because of the angle-angle theorem. We have the same angle at your eye, and we have parallel lines for the diameter of the tube and the field of vision.

Anna:	The lines are parallel because that's how we held the tube. And parallel lines make congruent angles.
Matt:	So, if you know the length of the tube and its diameter, you can use the proportion that Leo showed, $D_T/L_T = D_W/L_W$, to figure out the field of vision for any distance you plug in for L_W.
Anna:	Or you could use the length and diameter of the tube to get the slope of the line. Then write the equation of the line, and use that to solve for field of vision.
Ms. Van Ledjte:	What do others think of what Anna and Matt shared? How does it fit with what you've already seen or what your group came up with?
Julisa:	I think it fits really well with my explanation. The drawing shows what I was saying, that the bigger opening means you can see more through the tube. But I like it because it also explains why the variables are related in the way that Nick and Leo said they were.

Discussion of students' work on tasks 2 and 3

As the students considered the new data and possible justifications for the relationship that they found in task 3, the teacher's hints and suggestions guided some groups, but not all. Julisa's group, for one, did not use the hint that geometry might play a role in the justification of their conjecture about the variables in the Field of Vision problem. Instead, her group chose an algebraic approach and looked for a pattern among the set of equations (one equation for each different tube's data set). Julisa's explanation to the class showed not only how her group found the pattern, but also why the pattern made sense for the physical phenomenon being studied (wider opening means greater field of vision, thus greater slope of the equation). It wasn't until Matt and Anna shared their solution that Julisa was able to *reflect* on her justification and *make the connection* with a geometric representation. Leo attempted to incorporate the suggestion to find a geometric model and wisely chose similar triangles. However, he was not able to model the physical situation accurately, and his justification demonstrated that he was missing some connections. Table 1.4 shows the key elements of reasoning and sense making and the reasoning habits (NCTM 2009, pp. 9–10, 31, 55) illustrated by the students' work in tasks 2 and 3.

Table 1.4
Key Elements and Reasoning Habits Illustrated in Tasks 2 and 3

Key Elements of Reasoning and Sense Making with Algebra and Geometry

    Meaningful use of symbols

        Using variables to construct and interpret expressions

    Connecting algebra with geometry

        Representing algebraic situations geometrically

    Construction and evaluation of geometric arguments

        Developing and evaluating arguments about figures to make sense of a situation

    Geometric connections and modeling

        Using geometric ideas in real-world situations

Reasoning Habits

    Analyzing a problem

        Identifying relevant mathematical concepts and representations

        Applying previously learned concepts to new situations

Table 1.4—*Continued*

Implementing a strategy
Organizing the solution, including calculations and data displays
Making logical deductions by extending initial findings
Seeking and using connections
Connecting different domains and representations
Reflecting on a solution
Interpreting a solution and how it answers the problem
Generalizing a solution

The Field of Vision tasks span a range of mathematical concepts within a context that allows students to make connections across content areas while making sense of the problem situation. The teacher afforded students the time needed to explore and discuss issues of measurement and data collection, problem-solving strategies, and the development of mathematical arguments. The open-endedness of the tasks gave students flexibility and autonomy to make decisions. At the same time, the teacher made it clear that she expected students to justify their solutions. With this expectation, students demonstrated solutions, shared justifications, and argued for the validity of their justifications on the basis of connections to the physical model and the data collection process.

Conclusion

The problems that serve as examples in this chapter relate to geometric concepts of congruence and similarity, but they also require understanding of transformational geometry, measurement, and algebraic representations. The richness of these problems allows for exploration and discussion that necessitate reasoning and sense making. Each classroom episode showed the teacher as guiding students but not directly providing solution strategies.

For instance, in Mr. Lee's class, students worked on the Rotating Square problem without the teacher's help at the start. When they began sharing their initial ideas, Mr. Lee asked questions of all students to push them to think beyond the static models that they had drawn on their paper. He continually asked them to think about how their ideas might translate to a more general situation—random location for the rotating square.

The Field of Vision problem was very open-ended at the start, but the three tasks were structured to build on one another, thus giving students a chance to explore one data set before adding new variables and new data sets. The hint in task 3 about connecting the data sets to a geometric model was helpful when the students reached the point of developing a justification. Eventually, all students were able to make that connection.

In exploring both problems—Rotating Squares and Field of Vision—students were expected to work together to develop problem-solving skills, formulate conjectures to explain the mathematical situations, and use reasoning to explain or justify their results.

Chapter 2

Reasoning in Two Dimensions

Although our world is not flat, we typically represent it in two dimensions—consider, for example, primitive graphics on cave walls, drawings on paper, and images on our computer screens. We find comfort in being able to "contain" the world in which we live within the confines of a two-dimensional model. It sometimes seems easier to visualize and explain what happens on a flat surface than to venture into three-dimensional space. The geometry of the plane provides a stepping-stone into the real universe and gives us a way of organizing and simplifying ideas that might be daunting for the beginning intellectual explorer or uninitiated student.

It is important, though, that we challenge students to observe and explain what they see. Instead of defining a geometric shape for them, for example, encouraging them to invent their own definitions can be much more powerful, allowing them to take advantage of their innate inductive reasoning ability and use observed characteristics to come up with a generalized definition. Ownership of this process is guaranteed to generate interest and assures students that we are all, indeed, potential mathematicians, because we can all wonder about and process mathematical ideas.

The two sets of tasks presented in this chapter involve investigations of area formulas for certain polygons and explorations related to congruences of angles formed by diagonals of any polygon. The first set of tasks, Triangles and Quadrilaterals—Exploring Area Formulas, encourages students to find similarities among ways of determining the area of polygons. Although teachers may assume that high school students understand the concepts of area and perimeter, questioning them soon reveals, even in advanced classes, that much of the mathematical "success" that they have achieved is due, in large part, to their ability to memorize algorithms and definitions, with just enough understanding to apply some of the concepts. The tasks in this set can help students develop a deeper understanding of the meaning of area as they manipulate shapes or parts of shapes and see that the number assigned as the area does not change when the shapes are rotated, translated, or cut into smaller pieces. The responses and thinking processes of students in a class that implemented this set of tasks highlight the potential for student reasoning.

The second set of tasks, Angles and Diagonals, is presented through a series of possible, though fictitious, scenarios that suggest how students might discover properties of angles formed by diagonals of polygons. The congruence of these angles may not be intuitively obvious to students. It may take the teacher's guidance and careful exploration of angle measurements for students to produce a conjecture that will make sense to them or to discover that many aspects of the tasks connect to fundamental concepts of geometry, now presented in a "new" context.

Triangles and Quadrilaterals—Exploring Area Formulas

A ninth-grade geometry class is beginning a unit on measurement. The classroom teacher planned to introduce the unit by having the students work on a real-world problem related to tiling floors—the Creative Flooring problem, which appears below (see p. 33). Before sharing the problem, however, the teacher decides to review some of the concepts and formulas that the students will need in solving it.

Reviewing the idea of area

To review concepts related to area and how to determine it, the teacher introduces a simple tiling scenario:

Teacher: When we talk about tiling a floor, what type of measurements do we need to find?

Students: [*Sharing a variety of responses*] Area. Length. Width. Total surface area.

Teacher: So one thing we might want to know is the area of the floor. What does *area* mean to you?

[*Students, using squares and non-square rectangles, express their understanding that the area of any rectangle is the number of squares that can fit "inside" the rectangle, and that is why area is expressed in square units.*]

Teacher: Suppose we look at square tiles that are 1 foot by 1 foot. What is the area of this shape?

Students: [*Responding together*] One square foot.

Teacher: How would you know how many of these tiles are needed to tile a rectangular room that is 10 feet by 8 feet?

[*Students see easily that they would need 80 tiles because 10 multiplied by 8 is 80.*]

Teacher: What if your tiles are 2 feet by 2 feet? How many of these will you need to tile a room that is 10 feet by 8 feet?

Student 1:   We need to divide everything by 2. We only need half the number of tiles because they are twice as big.

Student 2:   Yeah, it'll take 40 tiles.

[*Several students nod their heads and mumble in agreement.*]

Student 3:   [*Working on something in her notebook before talking*] I don't think that's right. I drew a picture, and I think you need only 20 tiles, because the 2-by-2 tiles are twice as big in both directions.

Teacher: Why don't you show us your picture? Can you reproduce it on the board?

[*Student 3 draws her picture on the board, as shown in fig. 2.1, and explains her thinking to the class.*]

Student 3:   The dotted lines show that it's a 10-foot by 8-foot room, and the other lines show that you only need half as many tiles in each direction. I counted every other dotted line and got 5 by 4. So I figured you need only 20 tiles.

After a few more minutes of discussion, students come to the conclusion that they would need to divide 80 square feet (the area of the room floor) by 4 square feet (the area of each 2-by-2 tile) to get an answer of 20 tiles.

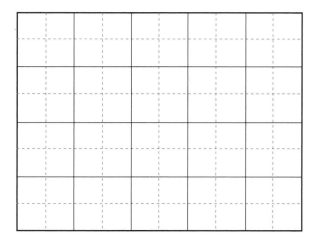

Fig. 2.1. Tiling an 8-foot by 10-foot room with tiles that
are 2 feet by 2 feet

In introducing a new topic or problem, it is always a good idea to assess students' prior knowledge and reasoning by leading a discussion about basic concepts. As illustrated by this introductory discussion of area, changing the question to consider 2-by-2 tiles instead of 1-by-1 tiles revealed that several students incorrectly assumed that doubling the dimensions of the square tiles would halve the number of tiles required, rather than divide it by 4. Asking students to share their definitions and their reasoning can provide a powerful assessment tool.

Exploring quadrilaterals

Next, the teacher turns the discussion to concepts related to classifying quadrilaterals before considering how to find the area of more general quadrilaterals. The following dialogue ensues:

Teacher: Now that you have a better sense about how area works, let's review our "most popular" quadrilaterals. We've studied parallelograms and trapezoids. Can you explain the differences between parallelograms and trapezoids?

Student 4:   A parallelogram has all sides parallel, but a trapezoid has one parallel side.

Student 5:   All sides of a trapezoid are different.

Teacher: Is it always true that all sides of a trapezoid have different lengths? Can trapezoids have sides with equal lengths?

Student 4:   No, because then it'll be a parallelogram.

Teacher: So, parallelograms have sides with equal lengths? And parallel sides?

Student 6:   I think you have to say that the opposite sides of a parallelogram are parallel, and they have equal lengths.

Student 7:   But a trapezoid can have opposite sides the same, too. Can I draw one? [*Draws a shape like that in fig. 2.2 on the board.*]

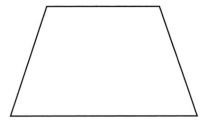

Fig. 2.2. A trapezoid with a pair of congruent opposite sides

Students continue to share ideas and build on the ideas of others. It takes several more attempts for them to differentiate the shapes and come up with a valid definition: A *parallelogram* is a quadrilateral that has opposite sides parallel (or opposite sides equal). A *trapezoid* is a quadrilateral that has only one set (or exactly one set) of parallel sides. The teacher explains that the two nonparallel sides of a trapezoid, called the *legs,* do not necessarily have the same length, and the students draw several examples of isosceles and non-isosceles trapezoids.

> Some consider parallelograms as a special subset of the class of trapezoid. That is, they allow trapezoids to have *at least one* set of parallel sides. The definition of trapezoid that the students come up with in the vignette is likely to be the most common or familiar. Clarifying a preferred definition will help students to communicate more effectively with one another.

The teacher follows up on this pair of definitions by drawing on the board a diagram that categorizes quadrilaterals as parallelograms, trapezoids, and others, and shows the relationships among them (see fig. 2.3). From parallelograms come two categories—rectangles and rhombi—and finally, another category of quadrilaterals—squares—joins these two types when rectangles are also rhombi. The classroom discussion continues.

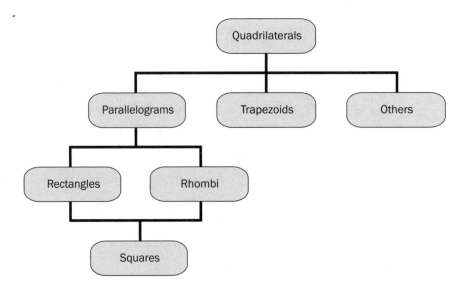

Fig. 2.3. A common classification of quadrilaterals

Teacher: Let's talk about the area of some of these figures. Let's start with a parallelogram that is not a rectangle [*draws such a figure on the board*]. Can we use our last definition of the area of a rectangle to determine the area of a parallelogram?

[*Students offer differing opinions. Some believe they can use the squares, while others believe that something else is needed.*]

Teacher: OK, how about drawing a parallelogram that is not a rectangle. We have centimeter grid paper. Draw a segment that is 4 centimeters long. Two centimeters above and one centimeter to the right, draw another segment, also 4 centimeters long [*illustrates on the overhead projector*]. Can you tell me how many square centimeters are in this parallelogram?

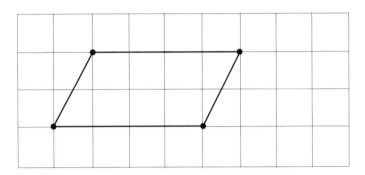

Fig. 2.4. A parallelogram on centimeter grid paper

Student 1: Some squares have corners sticking out.

Teacher: What does that mean?

Student 1: You can't use the 1-by-1 squares like we did for a rectangle. They don't fit nicely into the corners of the parallelogram. But you can estimate it. You can look at pieces of squares and try to match up the pieces inside the parallelogram to make whole squares.

Teacher: Can you tell what the area of that parallelogram is?

Student 2: I think it's 8 square centimeters.

Teacher: Why do you think that?

Student 2: There are 6 squares in the middle of the parallelogram. And if you count the squares that are not totally inside the shape as halves, you get 4 halves or 2 whole squares: 6 + 2 = 8.

Teacher: Can you do that?

Student 1: I did almost the same thing. I saw that 2 of the pieces of squares are more than a half and 2 are less than a half. When you put them together you get 2 wholes. I got 8, too.

[*Most students agree that this is a plausible method for finding the area of the parallelogram.*]

Teacher: What if we didn't have the grid behind the parallelogram to help us? Is there another way to find the area of a parallelogram?

The teacher distributes copies of a large parallelogram printed on plain paper and provides scissors and rulers to each student. The students work in pairs or groups of three to explore methods of finding the area of the parallelogram. The following conversation develops between students 3 and 4 as they work together.

Student 3:   Let's cut the parallelogram into pieces like this: go straight down here and up here [*draws with a pencil to indicate where she would cut, as shown in fig. 2.5a*]. When we cut these ends off, we get a rectangle. Then I think these two end pieces will make another rectangle. So we can use squares to find the area of the two rectangles.

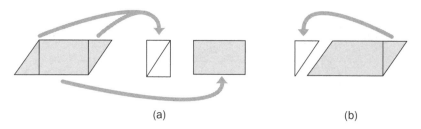

        (a)   (b)

Fig. 2.5. Dissecting a parallelogram to get a rectangle

Student 4:   We need to cut it only once. Cut it into two pieces, using one of your lines, like the one perpendicular to the top. Put this end piece at the other end of the parallelogram [*as illustrated in fig. 2.5b*]. You have a rectangle. See, right angles here and right angles here [*points to the four corners of the resulting figure*].

> Although in this scenario the students were able to come up with two valid methods for finding the area of the parallelogram, in other situations students might need guidance and encouragement from the teacher. For example, the teacher might ask students to think about deconstructing the parallelogram into objects for which they can easily find the area, such as squares and rectangles. The teacher could also suggest creating right angles by drawing perpendicular lines.

The teacher notices the work of students 3 and 4. She questions them:

Teacher: How did we say that we get the area of a rectangle?

Student 3:   We can find out how many squares of length 1 and width 1 fit inside.

Student 4:   Length multiplied by width. That's what we did to find the area of the room that was 10 feet by 8 feet.

Teacher: Can we say "base times height"?

Students: Isn't that the same?

Teacher: Yes, but by using *base* and *height,* it'll be easier to relate these measurements back to your original parallelogram.

At this point, the teacher calls for the attention of all students. Various pairs share what they did to find the area of the parallelogram. The teacher invites students 3 and 4 to the front of the room to summarize what they did, and the teacher repeats to the class her questions to these two students about using the terms *base* and *height* instead of *length* and *width* to describe how they would find the area of the rectangle created by transforming the parallelogram.

Teacher: What about for the parallelogram? Where would you say these terms apply on the figure?

Student 3:   We can say that the length of the rectangle that we created is the base of the parallelogram. They're the same thing, top and bottom. And the width of the rectangle that we created is the height of our parallelogram. I remember this from eighth grade.

As students continue to share thoughts about this, the teacher draws and labels a parallelogram on the board (see fig. 2.6) and also writes the familiar area formula:

$$\text{Area of parallelogram} = b \cdot h,$$

where *b* represents the length of one *base* of the parallelogram and *h* represents the perpendicular distance from one base to the other—the parallelogram's *height*.

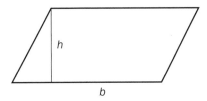

Fig. 2.6. Showing the dimensions of the
parallelogram

Deconstructing a parallelogram and making a rectangle is an important first step for understanding how to find the parallelogram's area. Making the connection between the original parallelogram and the resulting rectangle is an equally important component for making sense of the area formula for the parallelogram. Questioning students can help them make this connection.

To move the discussion in a new direction, the teacher suggests that students consider the area of a parallelogram as the average of the bases multiplied by the height.

Teacher: So, just for the sake of saying things in a different way, could you say that the area of a parallelogram with bases that are 4 centimeters and height that is 2 centimeters can be obtained by getting the average of the two bases—4 + 4, divided by 2, which is also 4—multiplied by the height, 2?

Student 5:   That's a strange way of doing it, but it would get the same result.

Student 6:   That doesn't make sense. The average of two numbers that are the same is just one of the numbers. It's like saying 2 times 4 divided by 2. It's just extra work. Why would you bother calculating the average of the bases? It's easier to use just one base to find the area.

Teacher: OK, but you'll see what I'm thinking about in a little while. Let's write this area formula on the board, at the top corner, so we don't erase it.

The students watch as the teacher writes,

Area of a parallelogram = $((b + b) \div 2) \cdot h$, where b is the length of the base and h represents the height of the parallelogram.

She then turns the discussion to finding the area of a trapezoid: "Suppose we look at trapezoids. Could we also apply the idea of area as the number of square units inside the shape?" The teacher distributes a handout showing an isosceles trapezoid that students can cut out and dissect. She asks, "Could we use similar techniques for deconstructing the trapezoid and reconstructing a different shape in order to find its area?" She invites the students to explore this idea.

Students 5 and 6 work together. They begin by cutting their trapezoids from the surrounding paper. Student 5 takes both trapezoids and fits them together as shown in figure 2.7. A discussion develops between the students, and eventually the teacher joins in:

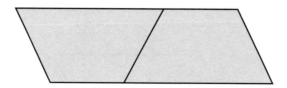

Fig. 2.7. A double trapezoid

Student 5: Well, it seems that if you turn one side around, they match up.

Student 6: What do you mean?

Student 5: Well, if I get my trapezoid, flip it like we saw others do with the pieces of the parallelogram, and put it next to your trapezoid, the sides match to make a parallelogram [*points to the pair of trapezoids on his desk*].

Student 6: OK, it does look like a parallelogram. How do we get the area of the trapezoid from this?

Student 5: The parallelogram is a "double trapezoid," so we just divide by 2.

Student 6: How do we get the area of the parallelogram?

Student 5: It's like she just showed us. Base times height.

Student 6: What's the height?

Student 5: It's this perpendicular line. Remember how she [*pointing to student 4*] showed that she cut her parallelogram to make a rectangle, and where she cut it made the height?

Student 6: OK, and the base is both of these lengths added together.

Teacher: [*Passing by, questions students 5 and 6 about what they have discovered*] Can you tell me the area of this "double trapezoid" parallelogram? Does that help me to get the area of one of the trapezoids?

Student 5:	Yes, because you put two of the same together to get the parallelogram, so if you cut it in two equal parts…
Teacher:	Tell me what you would have to do to get the area of one trapezoid.
Student 5:	Divide it by 2.
Teacher:	Can one of you walk us through the steps that it took to get this?

The teacher asks the class to stop work and listen to what students 5 and 6 have done. Student 5 tapes his "double trapezoid" to the front board and explains.

Student 5:	So if we add the two bases together and multiply by the height, that's the area of the parallelogram. And then that number can be divided by 2 to get the area of the trapezoid.

As before with the parallelogram, students working together were able to find a valid method for calculating the area of the trapezoid. The teacher, listening to two students' conversation, was able to question the students and help them to confirm their discovery. Summarizing their results, one of the students helped both students solidify the reasoning behind this approach. At the start of the activity, a teacher-led discussion focused on possible ways to deconstruct the trapezoid could have helped to encourage students to find other strategies.

Teacher:	I am writing this formula that you have discovered [*says the formula aloud while writing it at the top of the board*]: $((b_1 + b_2) \cdot h) \div 2$. Let's look at the other formula that we placed on the top right of the board [*reads out the earlier formula for the area of a parallelogram*]: $((b + b) \div 2) \cdot h$. Do you notice anything?
Student 3:	The b's are different.
Student 1:	One formula, you divide first, but the other one, you multiply times h first.
Teacher:	Let's address those two points. What do all of the b's stand for?
Several students:	The bases.
Teacher:	Why did I write little numbers at the bottom for the b's (the bases) of the trapezoid?
Several students:	Because they are different.
Teacher:	Do we agree that all the b's still stand for bases, although in the trapezoid, the bases are different, but in the parallelogram, the bases are the same?
[*Students agree.*]	
Teacher:	Let's now address the order of the multiplication and the division in the formulas. Think of a number. Use your calculator and multiply it by 6. Divide it by 4. Write your answer down. Now, take that same original number, divide it by 4, then multiply it by 6. Are the answers the same?
Students:	Yes.
Teacher:	Let's write this algebraically: If x is your number, $6x \div 4$ has the same value as $^x/_4 \cdot 6$. So your formula, $((b + b) \div 2) \cdot h$, can also be written as $(b + b) \cdot {}^h/_2$.

Can anyone now give me, in your own words, directions on how to find the area of a parallelogram or the area of a trapezoid?

The students have difficulty assigning words to the formulas. With prompting from the teacher, one student offers an answer:

Student 4: Can you say that you add the bases, divide that by 2, then multiply by the height? Or add the bases, multiply by the height and then divide that by 2?

Teacher: That's good. What are you doing when you add two quantities and then divide that sum by 2?

Several students: Getting an average.

Teacher: I now want you to restate what she [*indicates student 4*] said, using the word *average*.

Student 4: Average the bases, and then multiply by the height!

Teacher: That's excellent.

Discussion of students' reasoning in exploring area

Many other possible lines of reasoning might have developed in this classroom, leading to the area formula for the trapezoid. For instance, students could divide the trapezoid into a rectangle and two triangles and add the areas of these three figures. Drawing on prior knowledge about the area of triangles is essential. Students could then be asked to show that other versions of the trapezoid area formula are equivalent to the standard area formula:

$$\text{Area of a trapezoid} = (b_1 + b_2) \cdot \frac{h}{2}$$

Up to this point in the lesson, the classroom teacher worked with the students to review and strengthen their concept of area, with particular attention to the area formulas for quadrilaterals. She encouraged all students to use physical models, paper and pencil, as well as mental visualization to reason about the validity of the area formulas and the connections among them. Through her questioning and guidance, the teacher also helped students analyze the various geometric figures, look for relationships, and develop a generalized area formula that could work for either trapezoids or parallelograms. Table 2.1 identifies the key elements and reasoning habits (NCTM 2009, pp. 9–10, 55) illustrated in this classroom review of area formulas.

Table 2.1
Key Elements and Reasoning Habits Illustrated in Triangles and Quadrilaterals—Exploring Area Formulas: A Review

---

Key Elements of Reasoning and Sense Making with Geometry
  Conjecturing about geometric objects
    Analyzing shape relationships, such as the width of rectangle and height of parallelogram
  Constructing and evaluating geometric arguments
    Developing and evaluating arguments about properties of figures to make sense of geometric situations
  Using multiple geometric approaches
    Analyzing mathematical situations by using transformations, synthetic approaches, and spatial visualization

Table 2.1—*Continued*

Reasoning Habits

Analyzing a problem

Looking for hidden structure
- Drawing auxiliary lines on geometric figures
- Finding equivalent forms of expressions

Seeking patterns and relationships by carefully examining related figures

Implementing a strategy

Making logical deductions
- Verifying conjectures about area formulas
- Extending initial findings to find common area formulas

Seeking and using connections

Area formulas and the Creative Flooring Problem

The discussion in the ninth-grade geometry class changes direction again as the teacher introduces the problem that she originally planned to present, before deciding to prepare the students by reviewing:

Creative Flooring Problem

Martina Gonzalez, a retired mathematics teacher, and her husband own Creative Flooring, a ceramic tile company. Creative Flooring designs and produces tiles in the colors and dimensions that the client specifies, as long as the tiles are triangles, parallelograms (including rectangles that can be squares), or trapezoids. The company will use only one shape to tile each room. Martina would like to create a computer program that will quickly determine the number of tiles needed for a specific room when someone enters the dimensions of the desired tiles and the dimensions of the room to be tiled. She has a hunch that the program could rely on just one area formula for all the shapes that the company stocks.

Is there *one* formula that will work for triangles, parallelograms, and trapezoids?

Initial student reasoning

Again working in pairs, the class begins to think about the problem. Many students appear confused about how to begin answering the question posed by the task. After a short time, the teacher initiates a discussion by asking the following question: "Do you think it would be convenient to have only one formula to calculate the area of a tile no matter what shape it is? Why or why not?"

Students discuss the possibility that whoever is entering the information into the computer might not know the name of the shape being used or may make a mistake and enter the information into the wrong formula. One student suggests that the similar shapes but different area formulas for the trapezoid and parallelogram might confuse workers at Creative Flooring. All students agree that the work they've done on finding and writing a common formula for these two figures would be helpful.

Exploring triangles

The teacher reminds the students that the problem mentions triangles as a possible floor tile shape as well as parallelograms and trapezoids, and the following discussion develops:

Teacher: How do we define a triangle?

Student 7:   It has 180°.

Student 8:   It's a polygon with only three sides.

Teacher: Which description is more helpful for thinking about the area of a triangle?

Student 7:   Polygon with three sides. It tells you what the triangle looks like.

Teacher: So, how do we get the area of a triangle?

Students: [*Together, offering two choices*] Base times height. Base times height divided by 2.

Teacher: Let's remember what we did with the parallelograms and trapezoids [*redraws the figure shared earlier by student 5 (as shown in fig. 2.7)*].

Student 1:   Oh, we can do this with a triangle. Draw one triangle and one just like it but upside down. Can I show you on the board?

Teacher: Yes, draw a diagram of what you are thinking.

[*Student 1 draws two triangles as shown in fig. 2.8a on the board.*]

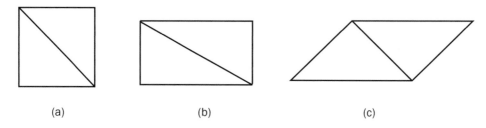

      (a)   (b)   (c)

Fig. 2.8. Two copies of three different triangles, with one copy rotated in each case to make a parallelogram

Teacher: Does it always happen that when you put a triangle and its rotated copy together you get a square?

[*Some discussion follows, and other students offer other illustrations similar to those shown in fig. 2.8b and fig. 2.8c.*]

Teacher: How does this resemble our double trapezoids above?

Student 1:   One shape is upside down.

Student 4:   You get a parallelogram when you put them together.

Teacher: Do we remember how we put the two trapezoids together to form a parallelogram, then used that to get the area of one of the trapezoids? How is that process similar to this one?

Student 3:   You can get the area of the parallelogram and cut it in half.

Teacher: So, what would be the process?

Several students: Base times height divided by 2.

Teacher: [*Pointing to the two area formulas still written on the board*] How is this different from the two formulas that we have discovered?

[*Several students point to the fact that a triangle has only one base, whereas the parallelogram and trapezoid each had two.*]

Teacher: To explore more similarities among the triangle, parallelogram, and trapezoid, we're going to stretch our concept of length a little. Could a segment have a length of 1 centimeter?

Students: Yes.

Teacher: How about $^1/_2$ centimeter?

Students: [*Again responding together*] Yes.

Teacher: How about $^1/_2$ of $^1/_2$? [*Continues, suggesting smaller and smaller numbers.*] How small could a segment be?

Students: [*Answering variously*] Tiny. Super small. Almost zero.

Teacher: Let's look at a trapezoid again. If we extended the legs upward in this case, what is happening to the top base? [*The teacher again draws an isosceles trapezoid and then makes a copy of it with extended legs, as in fig. 2.9.*]

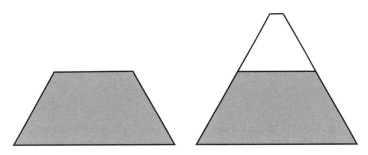

Fig. 2.9. Extending the legs of a trapezoid to decrease the length
of the top base

[*Students respond that the top base is getting shorter as the legs are extended upward.*]

Teacher: Can we still get the area of this trapezoid by adding the two bases, dividing by 2, and multiplying by the height?

Student 1: Yes, it would still be a trapezoid.

Teacher: What if we continue extending the legs? What happens if we extend them until they intersect? [*Makes a drawing like that in fig. 2.10.*]

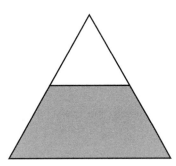

Fig. 2.10. The top base of the trapezoid approaches zero.

Student 2: It becomes a triangle.

Teacher: What has happened to the length of the top base?

Student 1:   It disappeared; it got too small.

Teacher: Can we say the length is zero?

[*The students agree that this is a possibility but express confusion about where this line of reasoning is going.*]

Teacher: OK [*moving on, hoping to be able to show the connections later, rather than tell about them now*]. We can think of that top base as having "collapsed" into a point, and we can say now that this triangle has two "parallel bases" but that one is really a point, a segment of length zero. Can you accept this?

[*The students discuss this question until they agree that this terminology does not contradict other concepts that they have learned about lengths, segments, shapes, and so on.*]

Although dynamic geometry software was not used during the discussion in this classroom, such software could be useful for demonstrating the shrinking top base of the trapezoid. The idea of a segment of length zero might be counterintuitive to students. In fact, segments are often defined to have distinct endpoints, and thus can never have a length of zero. In this classroom discussion, however, the teacher introduced the idea of a segment of length zero as a way to connect the area of a triangle with the formulas previously discussed for area of a parallelogram and trapezoid.

Because the connections among the area formulas are not usually obvious to students—particularly, the connection between the formulas for the areas of a triangle and a trapezoid—the teacher led this discussion much more actively, with much less exploration by the students.

Teacher: So, how can we then apply this idea to create a new formula that gives us the area of a triangle and connects to what we've been discussing for the areas of parallelograms and trapezoids? [*Draws on the board the diagram shown in fig. 2.11.*]

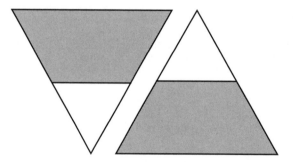

Fig. 2.11. Considering connections among the formulas for
the areas of a triangle, a trapezoid, and a parallelogram

[*Students start discussing in small groups. As the teacher listens to their conversations, she realizes that they still have some misunderstanding about the meaning of the variables in the formulas for the areas of a parallelogram and a trapezoid.*]

Teacher: Let's look again at the formula that we modified to obtain the area of a parallelogram, $((b + b) \div 2) \bullet h$. Why did we write b twice? When you first learned the formula for the area of a parallelogram, you learned only "base times height."

Student 3:	Oh, we did it so we could get a formula like the one for the trapezoid.
Student 4:	We could call the top base b_1 and the bottom base b_2, even if b_1 is equal to b_2. That way, it'll look just like the trapezoid formula.
Teacher:	So we have two different variables for the base, but in this case, $b_1 = b_2$. Can we do this now with the triangle? Can we label the "usual" base of the triangle as b_1 and the "collapsed" one, the one with length zero, b_2?
Student 5:	[*Sounding excited*] Sure! That way we can use the same formula as the trapezoid!

[*Other students begin to nod and show that they understand.*]

Teacher:	Then tell me how this formula, $((b_1 + b_2) \bullet h) \div 2$, could be applied to obtain the area of a triangle? What do you get?
Student 4:	Well, b_2 would be zero. So the formula is $(b_1 \bullet h) \div 2$.
Teacher:	[*To everyone*] Do you see the similarities? What I would like for you to do now, in your groups, is write an explanation of why the traditional formula used to get the area of a trapezoid (average of the bases times the height) can also be applied to get the areas of parallelograms and triangles. Then describe how this would help Creative Flooring. And to keep you thinking about this, consider another question: Could the company add other shapes to its menu of tiles and still use the same formula in its computer program?

Discussion of students' work in exploring area fromulas

The goal of Triangles and Quadrilaterals—Exploring Area Formulas is to help students see why the familiar area formulas work. In the vignettes, the teacher also introduced the students to the idea that the area formulas for the trapezoid, parallelogram, and triangle are connected. In fact, the class found a way to represent the areas of all three figures with just one formula. Although extending the trapezoid formula to include triangles was less intuitive to the students, they did eventually agree that this connection could be made.

The students spotlighted in the vignettes were enrolled in an honors ninth-grade geometry class. They were one year ahead of most of their peers, and had previously had one successful year of algebra. Yet, their teacher had to review many background concepts and address a few misconceptions to achieve the goal of the lesson. The activity can be used in a range of geometry and algebra classes—even in a more informal geometry class. In such a situation, however, more exploration of the basic concepts might be necessary, since students might take more time to arrive at the desired conclusion.

The lesson was taught in a 100-minute block, which facilitated the use of manipulative materials and the interaction of the students. In class periods of one hour or less, the lesson should be split, with most of the underlying concepts addressed on the first day. Thus, work on the second day would focus on "discovering" that it is possible to obtain the areas of all three shapes—parallelograms, trapezoids, and triangles—by using the same formula. Again, students in lower-level classes might need a longer period dedicated to understanding the "building block" concepts.

Students might use different tools to explore the concept of area and to discover the rationale for the area formulas of geometric shapes. Manipulating shapes on paper, creating shapes by using interactive geometry programs, sharing ideas with peers, and conjecturing on how the formulas relate are all ways for students to become familiar with the concepts in the lesson. Students frequently view formulas as mysterious equations created by mathematicians and beyond the realm of the average person's understanding. Bringing formulas to the students' level is an important step in sense making, thus building their confidence. Working through activities like Triangles and Quadrilaterals—

Exploring Area Formulas and this chapter's second set of tasks, Angles and Diagonals, makes geometry concepts interesting and real.

At the end of the lesson described here, the teacher encourages the students to write about what they have discovered and explain in their own words why the formulas work and how they are connected. This is an important follow-up activity for students, whether they do the writing independently or in groups. This task will help them think through each process and learn about how they actually *know* that something is true. What the students write can be used as a form of assessment, allowing the teacher to see the depth of the their understanding. Table 2.2 highlights the key elements and reasoning habits (NCTM 2009, pp. 9–10, 32, 55) illustrated by the students' work on the Creative Flooring problem, which concludes their work in Triangles and Quadrilaterals—Exploring Area Formulas.

Table 2.2
Key Elements and Reasoning Habits Illustrated in Triangles and Quadrilaterals—Exploring Area Formulas: Creative Flooring Problem

---

Key Elements of Reasoning and Sense Making with Geometry and Algebra

Conjecturing about geometric objects

Analyzing shape relationships such as the width of rectangle and height of parallelogram

Testing the validity of conjectures through inductive reasoning

Constructing and evaluating geometric arguments

Developing and evaluating arguments about properties of figures to make sense of geometric situations

Using multiple geometric approaches

Analyzing mathematical situations by using transformations, synthetic approaches, and spatial visualization as a way to develop and understand mathematical formulas

Connecting algebra with geometry

Representing geometric situations algebraically and algebraic situations geometrically, using connections in solving problems

---

Reasoning Habits

Analyzing a problem

Looking for hidden structure

• finding equivalent forms of expressions that reveal different aspects of a problem

Implementing a strategy

Organizing a solution

• formulating an algorithm

• using the structure of a problem

Reflecting on one's solution

Considering the reasonableness of a solution, including other ways of thinking about the problem

Generalizing a solution to a broader class of problems

Angles and Diagonals in Regular Polygons

The second set of tasks in this chapter is Angles and Diagonals. The problem, or motivating question, in this set is the following (fig. 2.12 illustrates the situation):

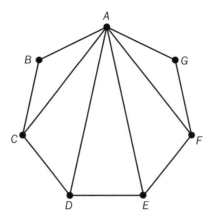

Fig. 2.12. A regular 7-sided polygon

Consider a regular heptagon and all the diagonals with endpoints at one vertex. Explain why the five adjacent angles with the vertex are congruent. Or, to be more precise and more general, for any regular polygon with n sides, show that all the diagonals with an endpoint at one vertex define congruent angles that measure $1/n$ times 180°.

For example, in figure 2.12, the five adjacent angles with vertex A are congruent, and each measures $(1/7)180°$.

> The mathematical and pedagogical goals of this set of tasks are less to demonstrate the stated proposition than to provide a situation that allows students to deepen and reinforce what they have learned about angles and polygons while practicing mathematical problem solving, reasoning, and communication. Such practice with a set of tasks that go beyond end-of-chapter exercises is valuable for students to consolidate concepts and increase flexible understanding.

In investigating this result, students explore intriguing angle relations in geometric figures, using some of what they have previously learned about angles. The ideas associated with this proposition set the stage for what students will learn when they later study inscribed angles in circles. (The reader may already have observed that this congruence of angles follows from the inscribed angle theorem, once the polygon is inscribed in a circle.)

What follows are descriptions of the tasks and possible classroom scenarios associated with them. Actual classroom dialogue is not included, but lines of questioning from the teacher (or students) and likely student reasoning in response to the teacher's questions are shared. The activities use a hands-on approach to engage students in investigating the situation described above. Possible student responses and reasoning related to the tasks are shown in indented text marked in the margin by gray triangles.

An experiment to set up the motivating question

The teacher gives the students some handouts showing regular polygons, with vertices labeled A, B, C, and so forth, and asks the students to draw all the diagonals containing endpoint A. This work dissects each polygon into triangles, as illustrated in figure 2.12. The teacher asks which angles they think are bigger, and some students see certain angles as different, but others claim that they are the same.

The students are then instructed to cut out the triangles and compare the angles that contain vertex *A*. They observe that the angles are congruent, even though the lengths of the segments are not equal. The experiment raises a question that leads into the rest of this set of tasks:

> Are these angles congruent for any regular *n*-gon? If so, why is this true? And what is the measure of these equal angles?

Cutting out shapes, even at the high school level, can help students isolate parts of complicated figures.

Initial classroom examination of angles and polygons

The teacher asks the students to recall facts, first about angle sums in triangles, and then about angle sums in quadrilaterals.

▶ All the students remember that the sum of the angle measures of a triangle equals 180 degrees. Most
▶ students remember that the sum of the angle measures of a quadrilateral equals 360 degrees, but no
▶ one is quite sure why.

The teacher draws a quadrilateral on the board and divides it into two triangles by a diagonal. The angle measures are labeled *a, b, c, d, e,* and *f,* as in figure 2.13.

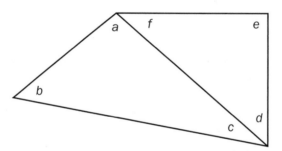

Fig. 2.13. A quadrilateral divided into two triangles
with angle measures labeled

The teacher instructs the students to draw a quadrilateral and divide and label it in this manner. They then (1) discuss in their groups how to write the sum of the angle measures of each triangle, using these letters; (2) write the sums; and (3) write the sum of the angle measures of the quadrilateral, using the same letters to see how the sums are related. One group volunteers to share what they have found and writes the following on the board:

▶ $a + b + c = 180°$
▶ $d + e + f = 180°$
▶ $b + (c + d) + e + (a + f) = (a + b + c) + (d + e + f) = 180° + 180° = 360°$

After the groups have found and discussed this relationship, the teacher gives each group a handout showing a regular pentagon and asks the students to find the angle sum in this case.

▶ All groups immediately divide the pentagon into triangles or quadrilaterals (two possibilities are
▶ shown in fig. 2.14). Most groups create a figure similar to figure 2.14a and calculate the sum of the
▶ angles of the 3 triangles. A few groups use a figure similar to figure 2.14b. One group uses variables
▶ to represent the angle measures, as in the work shown on the board related to the quadrilateral (fig.
▶ 2.13). This group's work progresses slowly.

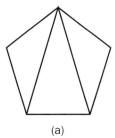
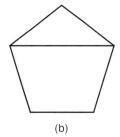

(a)　　　　　　　　　　　(b)

Figure 2.14. Dividing a regular pentagon into triangles

> Reviewing concepts by asking students to recall what they know rather than just explaining to them what they need to remember has at least two advantages. It involves students in the review, and it helps the teacher assess what they really remember and what needs reinforcement.

After most groups have arrived at the correct answer, determining that the angle sum of a regular pentagon is 540°, the teacher brings the class together. She calls on two groups to give their answer and explain their reasoning:

▶ • A member of the first group explains that they divided the pentagon into three triangles by drawing
▶ two diagonals. Then they added up 180 + 180 + 180 because each triangle has an angle sum of 180.
▶ A second member of this group adds that the angles of the triangle can be put together to make the
▶ angles of the pentagon.
▶
▶ • The second group's representative explains that their method was quicker because they knew the angle
▶ sum of the quadrilateral was 360°. They added 180 + 360 to get the answer of 540°.

The teacher then draws a convex hexagon and a convex heptagon on the board and asks for ideas about the angle sum. The students offer various suggestions:

▶ Several students suggest dividing the polygons into triangles and then suggest that the sum is equal
▶ in each case to the number of triangles times 180 degrees. One student asks if the triangles can be
▶ formed in any way or only by using diagonals from the same vertex.

The teacher turns this question back to the students. After more discussion and examples drawn on the board (see fig. 2.15), the class concludes that only diagonals will work because then the angles of the triangles are parts of the angles of the polygon that is being subdivided. Although the students agree that any diagonals will work, they also agree that for consistency, they will use just one vertex in each polygon to draw all diagonals (as in fig. 2.15c).

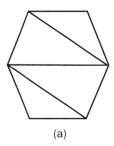 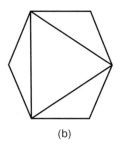 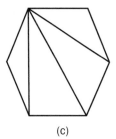

(a)   (b)   (c)

Figure 2.15. Three possible subdivisions of convex hexagon
into triangles

The teacher continues the discussion to guide the students in reasoning about a general rule for any (convex) polygon. She records information in a table, leaving a couple of columns blank, as shown in figure 2.16.

Number of Sides	Sum of Angle Measures		
3	$1 \cdot 180° = 180°$		
4	$2 \cdot 180° = 360°$		
5	$3 \cdot 180° = 540°$		
6	$4 \cdot 180° = 720°$		
7	$5 \cdot 180° = 900°$		
8	$6 \cdot 180° = 1080°$		
N	$(N-2) \cdot 180°$		

Fig. 2.16. Making a table to organize information about convex polygons

The use of tables for organizing information is very valuable for students. It is worthwhile to let students decide how to organize the tables. But the teacher may have to add or modify their suggestions to meet the goals of the lesson. In this case, the teacher made sure that the angle measures were written as multiples of 180 degrees and not just total degrees.

Exploring vertex angles of regular polygons

Next, the teacher states the definition of a *regular polygon* as a polygon with congruent sides and angles. Some regular polygons have already appeared in examples. The teacher asks, "If in a regular pentagon, the angle at a vertex is x, what is the value of x?" The students agree that the angle sum is $5x$, which equals 540 degrees. So the class concludes that the angle measure, x, is $\left(\frac{540}{5}\right)$, or 108 degrees.

The teacher now returns to the table made earlier (see fig. 2.16). She labels column 3 as "Angle Measure of a Vertex in a Regular Polygon" and gives the groups the task of filling in this column.

After a bit of discussion and struggle, the students complete the new column and talk about their work:

▶ Students have written the vertex angle measures in two ways. Everyone agrees that they can simply
▶ take the angle sum from the second column and divide that value by the number of sides (or angles)
▶ to get the measure of one vertex of the regular polygon. This gives $((n-2) \cdot 180°)/n$. But one student
▶ noticed that this expression could be rearranged because multiplication and division are commutative.
▶ He suggests keeping the variables together in the expression to find $(n-2)/n$ and then multiplying
▶ by 180°.

Both methods are recorded in the table in figure 2.17.

Number of Sides	Sum of Angle Measures	Angle Measure of a Vertex in a Regular Polygon	
3	$1 \cdot 180° = 180°$	$(1 \cdot 180°)/3 = (1/3) \cdot 180° = 60°$	
4	$2 \cdot 180° = 360°$	$(2 \cdot 180°)/4 = (2/4) \cdot 180° = 90°$	
5	$3 \cdot 180° = 540°$	$(3 \cdot 180°)/5 = (3/5) \cdot 180° = 108°$	
6	$4 \cdot 180° = 720°$	$(4 \cdot 180°)/6 = (4/6) \cdot 180° = 120°$	
7	$5 \cdot 180° = 900°$	$(5 \cdot 180°)/7 = (5/7) \cdot 180° \approx 128.57°$	
8	$6 \cdot 180° = 1080°$	$(6 \cdot 180°)/8 = (6/8) \cdot 180° = 135°$	
N	$(N-2) \cdot 180°$	$(N-2) \cdot 180°/N = (N-2)/N \cdot 180°$	

Fig. 2.17. Continuing the table shown in fig. 2.16

The teacher notes that the students will complete column 4—the final column in the table—by filling in the measures of the *exterior angles* of each regular polygon. To understand this concept and find this measure, the students undertake the following investigation with scissors.

A scissors investigation of exterior angles

Students are given a handout showing a copy of a regular pentagon. Because the term *exterior angle* is new to most of them, the teacher defines it:

> An *exterior angle* is the angle created by one side of a convex polygon at a vertex and the ray extending the adjacent side.

She demonstrates with a drawing on the board, as in figure 2.18.

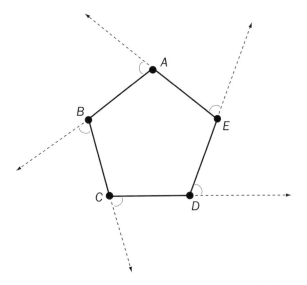

Fig. 2.18. A regular pentagon with exterior angles drawn

The students pose questions about this new concept. Some ask why the entire outside of the interior angle is not called the exterior angle. The teacher explains that the angle shown in the drawing is a very useful angle and is less than a straight angle, so geometers have given this angle the name of *exterior angle*. With guidance from the teacher, the students investigate an example and see that there are two exterior angles at any vertex, but the two angles are congruent vertical angles, so their measures are the same.

To begin the investigation of the measures of exterior angles of regular polygons, students draw and cut out the five exterior angles in a regular pentagon on the handout (as shown in fig. 2.18). They then manipulate them to answer the following questions:

- What is the relationship between any two of the exterior angles?
- What is the sum of the measures of the five exterior angles?
- Why do your answers to the first two questions make sense?

The students discover that the angles are all congruent and fit together to fill 360°.

The teacher then gives the students a handout showing several regular polygons: an equilateral triangle, a square, a regular hexagon, a regular heptagon, and a regular octagon. She instructs them to (1) draw exterior angles (as in fig. 2.18); (2) cut them out; and (3) for each regular polygon, find the relationship between any two exterior angles; and (4) find the sum of the exterior angles. The students again discover that the angles are all congruent and fit together to fill 360 degrees and thus have the same sum of 360 degrees for all of the polygons.

> This chapter uses the names of polygons, such as *heptagon, nonagon,* and so on. The mathematics works just as well if the class uses the names *7-gon, 9-gon,* and so forth. The correct name of a polygon with *n* sides is *n-gon*, anyway. Some students particularly enjoy learning the names with Greek stems, but for others these names introduce needless confusion about terminology. The teacher should make a judgment about the extent to which to introduce these terms in a particular class. Using both kinds of terminology together is possible and natural in some instances.

Sense making about the sum of exterior angles

The scissors experiment for the sum of exterior angles is convincing as evidence, but it offers no explanation of why the sum is always 360 degrees. In a whole-class discussion, the students work together with the teacher to raise the following points to help make sense of this relationship:

- The congruence of the exterior angles makes sense because the rotational symmetry of the figure takes one angle to another.

- The congruence of the exterior angles makes sense because each exterior angle is supplementary to an interior angle, and the interior angles are congruent.

- The scissors experiment can be performed on a regular *n*-gon in a different way: someone can start with the (cut-out) exterior angles in position and then slide them carefully to the center (without rotating). Then the angles are formed by *n* rays with a common vertex, arranged with the same rotational symmetry as the *n*-gon, since the rays are parallel to the sides of the *n*-gon.

- With interactive software, someone can start with a regular polygon (like that shown in fig. 2.18) and dilate the polygon to a very small size so that it looks like a point, while the exterior rays (being of infinite length) remain rays. But the figure on the screen now appears to be made of *n* rays arranged symmetrically around a common vertex. Because the shrinking does not change the exterior angles of the polygon, these angles must add up to 360 degrees.

Making sense of exterior angles with algebra

A more algebraic explanation comes from the table, the final column of which the teachers and students now fill in with the measures of the exterior angles (see fig. 2.19). The students reason as follows:

▶   Because the exterior angles are supplementary to the interior angles, it is possible to compute them. In
▶   each case, for a regular *n*-gon, the exterior angle measures (2/*n*) • 180 degrees. So the total is *n* times
▶   this number, or 2 • 180 = 360°.

The numerical examples in the table are easy to check. The last row, for *n* sides, may be a little hard for students to see, but by writing

$$\frac{n-2}{n} = \frac{n}{n} - \frac{2}{n} = 1 - \frac{2}{n},$$

the teacher can make it easier to understand.

Number of Sides	Sum of Angle Measures	Angle Measure of a Vertex in a Regular Polygon	Exterior Angle Measure in a Regular Polygon
3	1 • 180° = 180°	(1 • 180°)/3 = (1/3) • 180° = 60°	(2/3) • 180° = 120°
4	2 • 180° = 360°	(2 • 180°)/4 = (2/4) • 180° = 90°	(2/4) • 180° = 90°
5	3 • 180° = 540°	(3 • 180°)/5 = (3/5) • 180° = 108°	(2/5) • 180° = 72°
6	4 • 180° = 720°	(4 • 180°)/6 = (4/6) • 180° = 120°	(2/6) • 180° = 60°
7	5 • 180° = 900°	(5 • 180°)/7 = (5/7) • 180° ≈ 128.57°	(2/7) • 180° ≈ 51.43°
8	6 • 180° = 1080°	(6 • 180°)/8 = (6/8) • 180° = 135°	(2/8) • 180° = 45°
N	(*N* − 2) • 180°	(*N* − 2) • 180°/*N* = (*N* − 2)/*N* • 180°	(2/*N*) • 180°

Fig. 2.19. Completing the table from figs. 2.16 and 2.17

The formulas and patterns for interior and exterior angle measure look different, depending on whether they are written as fractions times 180° or as (sometimes approximate) decimal numbers. Although the decimal numbers make it clearer how large the angle is, it is easier to make sense of the formulas when the answers are represented by exact fractions. Thus, students should be encouraged at this point to maintain the fractions.

Exploring angles formed by the diagonals of a regular pentagon

The teacher gives the students another handout showing a regular pentagon. She asks them to label the vertices as *A–E* and draw all the diagonals with endpoints at vertex *A,* as in figure 2.20a.

This time, the task for students is to write angle measures next to each of the angles in the pentagonal figure, as in figure 2.20b. Since in this case two of the three triangles are congruent isosceles triangles, students can start with known angles and use the triangle angle sum to find the rest of the angles.

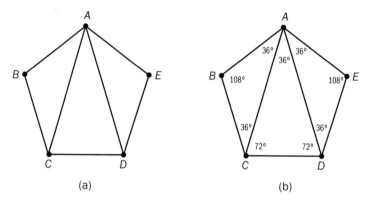

Fig. 2.20. A dissected regular pentagon and its angle measures

Class discussion of angle measures

In a follow-up class discussion, students share a variety of ways to find and justify the measures of the angles:

▶ • One student reasons that triangle ABC is isosceles, with angle $ABC = 108°$, since this is an interior
▶ angle of the pentagon. Thus, each of the congruent base angles, BAC and BCA, has a measure of
▶ $(180 – 108)/2 = 36$ degrees. The same quantities appear in triangle AED.
▶
▶ • Another student notes that the remaining angle measures could then be found by subtraction of an-
▶ gles. For example,

$$m\angle CAD = m\angle BAE – m\angle BAC – m\angle DAE = 108 – 36 – 36 = 36°.$$

▶ • Similarly, the measures of each of the angles ACD and ADC can be computed as $108 – 36$, or $72°$.

Another approach would be to notice that line CA is parallel to line DE, and line CD is a transversal. Thus, angle ACD must be congruent to an exterior angle of the pentagon at D, previously calculated to be 72 degrees. Thus, the students can compute the measures of all the angles in the pentagonal figure.

Angle measures in a regular heptagon

When the students have succeeded in the case of the regular pentagon, the teacher asks them to use the same process in considering a regular heptagon (see fig. 2.21). They dissect it in the same way (with diagonals from a single vertex) and identify as many of the angle measures in the resulting triangles as possible.

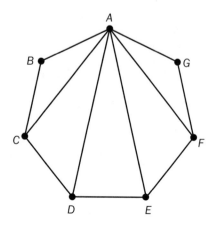

Fig. 2.21. A dissected regular heptagon

Some students may be confused about why they cannot just find the measures within the dissected heptagon (as shown in fig. 2.21) by assuming that all the angles at *A* are congruent, since they observed the usefulness of this assumption in the case of the pentagon. This misconception provides an opportunity to review the difference between a conjecture strongly supported by inductive evidence and a conclusion explained by deductive reasoning.

Although it is likely that students can figure out the entire pentagon case (perhaps with some hints), the case of the regular heptagon is not so simple. Referring to figure 2.21, note that the measures of the angles of triangles *ABC* and *AGF* can be found in the same way as for the pentagon, but triangle *ACD* presents a new problem. Although students can use the same method as before to find the measure of angle *DCA*, the measures of the two other angles of the triangle remain unknown. The heptagon case serves as a lead-in to the general problem. It is not necessary for students to solve the problem at this point since the next part of the exploration presents a general strategy.

Using a straight angle as a unit of measure

To discourage students from writing decimal approximations for the angle measures in a regular *n*-gon, the teacher instructs them to use as their unit of angle measure 1 straight angle, which will be written as 1 S in the following activities. Using this unit notation, the measure of an interior angle of a regular heptagon can be expressed as $\left(\frac{5}{7}\right)$S.

This discussion of an alternative unit has value for students in two ways. First, they can see that an appropriate choice of units can make a mathematical pattern more evident. Second, the translation from straight angle measure to degrees is good practice. It resembles closely the translation from radians to degrees.

Exploring angles in arches

With guidance from the teacher, students can now discover that a key to one approach to finding the measures of angles in regular polygons, such as the regular heptagon, is to find the angle measure in the basic case of the angle formed by a side of the polygon and any one of the diagonals. To do this, students consider a regular polygon with one diagonal drawn, as illustrated by two examples in figure 2.22.

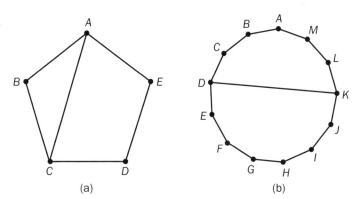

Fig. 2.22. Examples of regular polygons with one diagonal drawn

The diagonal separates the sides of the polygon into two sets of sides. Each of these sets, along with the diagonal, is a polygon itself. In this exploration, such a polygon will be defined as an *arch* of the regular polygon. Thus, any diagonal is the side of two arches of the polygon. In the regular pentagon in figure 2.22a, both triangle *CAB* and quadrilateral *ACDE* are arches, the first with two of the sides of the regular pentagon, and the second with three. In the regular 13-gon in figure 2.22b, the heptagon *KLMABCD* is an arch containing six sides of the regular polygon. The other arch has seven sides of the regular polygon.

The students' investigation of arches consists of finding angles in figures created by parallel diagonals, as in figures 2.24, 2.26 and 2.27. In these instances, drawing parallel diagonals creates triangles and trapezoids (see the note below about the inclusive definition of *trapezoid*). Thus, before the students begin, the teacher spends time drawing figures and reviewing with students some of what they know about trapezoids and parallel lines. The students are eager to share their understanding; their observations refer to an isosceles trapezoid labeled as in figure 2.23:

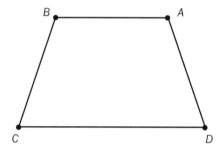

Fig. 2.23. A trapezoid with parallel sides *BA* and *CD*

- • If *ABCD* is a trapezoid with side *BA* parallel to side *CD*, then the interior vertex angles at the endpoints of side *BC* are supplementary (with the same relation for the angles at the endpoints of side *DA*).

- • If *ABCD* is a trapezoid with side *BA* parallel to side *CD*, then the interior vertex angle at *B* and the exterior angle at *C* are congruent (with the same relation for the interior vertex angle at *A* and the exterior angle at *D*).

- • If *ABCD* is a quadrilateral with angle *DAB* congruent to angle *CBA* and also side *AD* congruent to side *BC*, then *ABCD* is an isosceles trapezoid (including the case of a rectangle as a trapezoid).

In addition to understanding the idea of an arch by using visual examples, students are asked to write a definition of this new term. They do this in groups and come to a consensus as a whole class that an arch is a polygon formed within a larger polygon by drawing a diagonal. Students are also encouraged to think about how this concept will help them develop the reasoning to explain why the angles formed by diagonals at one vertex of a regular polygon are congruent.

As noted earlier, two definitions of *trapezoid* are current in mathematical textbooks. One definition states that a trapezoid is a quadrilateral with *exactly* one pair of parallel sides. This is the traditional, *exclusive* definition, which rules out parallelograms, and thus rectangles, as trapezoids. The second definition states that a trapezoid is a quadrilateral with *at least* one pair of parallel sides. This is the newer, *inclusive* definition, which categorizes parallelograms, and thus rectangles, as trapezoids. The inclusive definition works well in the context of this problem. Each definition has its advantages, but in a geometry class, using one definition consistently throughout the course makes sense.

Exploring nested arches and trapezoids: The heptagonal case

The students are now ready for the key arch activity. Before beginning this activity, the teacher reminds them to use as their unit of angle measure 1 straight angle, or 1 S, instead of degrees. For example, an interior angle of a square is $^1/_2$ of 1 S. Also, they are told to use fractions instead of decimals from a calculator, so that patterns will be more obvious.

Students are given a handout showing a heptagon with one pair of parallel diagonals, as shown in figure 2.24, and are asked to work in their groups to find the measure of each angle (in straight angle units) and to explain their reasoning.

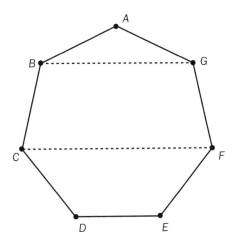

Fig. 2.24. A regular heptagon with parallel diagonals

The heptagonal case: A debriefing

Because the case under consideration is fairly basic, the teacher quickly debriefs the students before moving on to the next case. The students compare results and come to a consensus on the labeling of the angle measures (each fraction is a multiple of 1 straight angle), as shown in figure 2.25. The teacher draws a figure on the board, and students come up in turn to explain the reasoning used to find the angle measures.

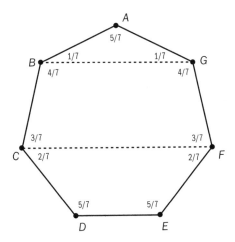

Fig. 2.25. Angle measures (in straight angle units) in the case of a
regular heptagon

▶ • All the students noticed that triangle ABG is isosceles. They also knew that the measure of angle BAG = $(5/7)S$ by the formula for the interior angle of a heptagon developed earlier (see fig. 2.17). Then, using the angle sum for triangles and the congruent base angles, they found that

$$2(m\angle ABG) = 1S - (5/7)S = (2/7)S,$$

so angle ABG and angle AGB each measure $(1/7)S$.

• The students noticed that quadrilateral $CFGB$ is an isosceles trapezoid. One student reasoned that the heptagon itself has a line of symmetry through vertex A and the midpoint of DE, and this line is also a line of symmetry for quadrilateral $CFGB$. But most students computed angle measures to see that $m\angle CBG = m\angle FGB$.

• The students found the measure of angle CBG by noticing that angle CBA is the sum of the measures of angle CBG and angle GBA. Thus, $m\angle CBG + (1/7)S = (5/7)S$, so angle CBG measures $(4/7)S$. Using the same reasoning also gives $m\angle FGB = (4/7)S$.

• Another student shared the observation that this means that angle CBG and angle FCB are supplementary, so $m\angle FCB = 1S - (4/7)S = (3/7)S$, which is also the measure of angle CFG.

• One student surprised the others by finding the measure of angle FCB in a different way. She extended the segment CB as a ray to form an exterior angle of the heptagon at B; this exterior angle has a measure of $(2/7)S$ (see fig. 2.19). Then the exterior angle of the trapezoid $CFGB$ at B is the sum of the measures of this angle and angle GBA, so the exterior angle of the trapezoid $CFGB$ measures $(2/7)S + (1/7)S = (3/7)S$, and this angle is congruent to angle FCB. (This is a case of corresponding angles formed by a transversal, BC, intersecting two parallel lines, BG and CF.)

• The last angle measures come from quadrilateral $DEFC$. Some students computed the measurements of the angles at C and F as before, as a difference of angle measures. But other students pointed out that they knew that the measures of the angles at D and E equal $(5/7)S$ since they are interior angles of the heptagon. This is enough to show that $DEFC$ is an isosceles trapezoid, and hence angle DCF and angle EFC are congruent and measure $(2/7)S$ (because they are supplementary to the angles at D and E).

• At the end of the discussion, the teacher guides the students in noticing some patterns. For example, $1/7, 3/7, 5/7$ (of a straight angle) are, respectively, the measures of the base angles of the two "upper" arches that contain vertex A (namely, triangle ABG and pentagon $CBAGF$) and the angle at vertex A. In addition, the base angles in the "lower" arches that contain segment DE measure $4/7$ and $2/7$. Thus, all the angle measures, $1/7, 2/7, 3/7, 4/7, 5/7$ are represented. Any angle between a side and a diagonal of the heptagon is congruent to one of these angles.

Displaying the results in a table

The students collect their results in a table (see table 2.3). The first column shows the number of sides of the original heptagon that are contained in the arch. The second column shows the measure of one of the base angles of the arch (in straight angle units). For example, triangle ABG is an arch that contains only two sides of the original heptagon. The base angles of this isosceles triangle measure $(1/7)S$. Pentagon $CBAGF$ is an arch that contains 4 sides of the original heptagon. Each of its base angles measures $(3/7)S$ (refer to fig. 2.25).

Table 2.3
Base Angles of Arches in a Regular Heptagon

Number of sides of the regular polygon that are sides of the arch	Measure of a base angle of the arch in straight angle units
2	$1/_7$
3	$2/_7$
4	$3/_7$
5	$4/_7$

Students have explored many ideas about angles and polygons by this point. Where teachers take the investigation next will depend on their students. With some classes, the teacher may judge that students will make the best use of time by working out special cases thoroughly and explaining them clearly. With other classes or groups of students, the teacher may decide that the students can move to generalization and a more algebraic formulation of what they have learned. No matter what direction students take, and however they use the ideas, they can benefit from reasoning about and explaining interesting examples such as these.

Extending to more polygon cases

The teacher gives the students several more examples of regular polygons with parallel diagonals, such as the regular 13-gon in figure 2.26. Again, the students label the measures of the angles by using the methods developed in the first example. Once again, they make tables of their measurement and begin to look for patterns and generalizations.

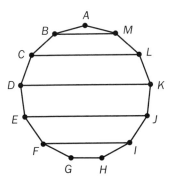

Fig. 2.26. Regular 13-gon with parallel diagonals

Polygons with an even number of sides present a slight complication: not all possible arches and their angles appear in a simple dissection by a set of parallel diagonals. For example, in figure 2.27a, the decagon has arches with only 2, 4, 6, and 8 decagon sides, so a table based on this dissection would be missing about half the cases. Figure 2.27b shows the same decagon in a different dissection by a set of parallel diagonals that form the arches with an odd number of sides from the regular decagon.

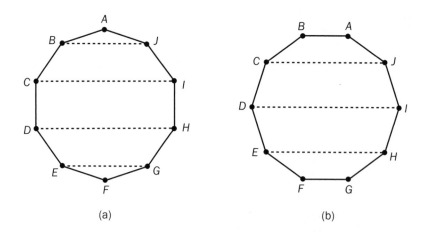

(a)   (b)

Fig. 2.27. A regular decagon with parallel diagonals

Figure 2.28 shows angle measures (in straight angle units) for the decagon in figure 2.27a. Table 2.4 shows measurements for all the arches of the regular decagon, combining the angle measures from parts (a) and (b) of figure 2.27.

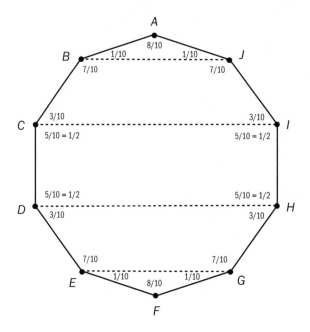

Fig. 2.28. Angle measures (in straight angle units) for fig. 2.27a

Table 2.4
Base Angles of Arches in the Regular Decagon (10-gon)

Number of sides of the regular polygon that are sides of the arch	Measure of a base angle of the arch in straight angle units
2	$^1/_{10}$
3	$^2/_{10}$
4	$^3/_{10}$
5	$^4/_{10}$
6	$^5/_{10}$
7	$^6/_{10}$
8	$^7/_{10}$

Summarizing the general pattern for arches

Students have a lot to summarize after investigating the arches and their angle measures. With several tables to look at, they can see that for an arch in a regular n-gon, if the arch is made from $(p + 1)$ sides of the polygon, then the base angle of the arch measures $(^p/_n) \times (1$ straight angle$)$, or $(^p/_n)$S.

> Observing patterns in tables is a valuable way for students to discover mathematical relationships. However, it is important for students to realize that assuming that a pattern recognized in a table is always true is akin to measuring two angles and concluding that they are congruent because the measurements seem to be the same. Both are cases of strong empirical evidence, but such observations do not establish that the conjecture is true. More important, empirical evidence does not explain why the pattern or conjecture is true.

Students should be able to justify this pattern on the basis of the figures with parallel diagonals. When moving from one arch to the next with two more sides, the measure of the base angle of the larger arch always increases by $(^2/_n)$S.

But one aspect of the pattern found in the table may still be troublesome to students. The relationship of arches with $p + 1$ sides to a measure of $(^p/_n)$S is a little awkward. Students may notice that p is the number of vertices in the arch minus the number of base vertices. Maybe this would be a better way to describe the arch and connect it to the measure of the base angles of the arch.

Final steps: Solution of the problem, and a generalization

For the last stage of the investigation, the teacher distributes handouts showing a number of figures, each consisting of a regular polygon and an angle PQR, where P, Q, and R are vertices of the polygon (but not necessarily consecutive). She asks the students to find the measure of the angle by using what they know about base angles for arches. For example, the measure of angle EAF in the 13-gon in figure 2.29a can be determined by using base angles of arches in several ways. One is to notice that the angle measure of EAF is the difference of the measures of angle BAF and angle BAE. Both of these angles are base angles of arches and hence their measurements are known; one is $(^4/_{13})$S, and the other is $(^3/_{13})$S, so the difference is $(^1/_{13})$S.

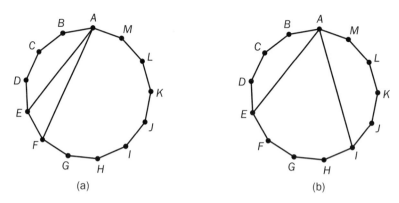

Fig. 2.29. Finding angle measures in the regular 13-gon

A second method is to consider the measure of the vertex angle *BAM* as the sum of the measures of angles *BAE, EAF,* and *FAM.* Two of these are base angles for arches, and the measure of *BAM* is $(^{11}/_{13})$S. Hence, we can write $(^{11}/_{13})$S = $(^3/_{13})$S + $m\angle EAF$ + $(^7/_{13})$S. Thus, the measure of angle *EAF* is $(^1/_{13})$S.

In figure 2.29b, students may notice how the angle measure changes when one of the diagonals is changed. The base angle *BAE* of the arch formed by pentagon *ABCDE* measures $(^3/_{13})$S. The arch with the base angle *BAI* contains 4 more sides. From the general pattern, the students know that the size of that base angle is larger by $(^4/_{13})$S. So angle *EAI* = $(^4/_{13})$S .

Students now return to the original regular heptagon (shown again for the readers' convenience as fig. 2.30) to analyze the measures of the angles created by drawing all diagonals from vertex *A*. In this case, they can now see that the diagonals with vertex *A* form five angles, each measuring $^1/_7$ of a straight angle (calculated by using methods described above for the 13-gon) and together adding up to $^5/_7$ of a straight angle. This analysis can be extended to any *n*-gon to show that for any regular polygon with *n* sides, all the diagonals from one particular vertex form $(n-2)$ angles, each of which measures $^1/_n$ × (1 straight angle).

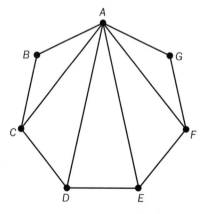

Fig. 2.30. Angles in the regular heptagon

The results of the students' investigations can be expressed in a general statement, which the teacher can share with them after they write their own formulations:

> **Theorem:** In a regular polygon with *n* sides, if *A* is a vertex and *P* and *Q* are other vertices, then the measure of the angle *PAQ* = $^k/_n$ × (1 straight angle) = $(^k/_n)$180 degrees, where *k* is the number of sides of the polygon that lie in the interior of the angle.

Students will understand and value statements that they generate themselves. In the ideal case, their formulations are complete and correct. But even if students must be supplied with a correct and polished version, they will appreciate it more and understand it better if they have struggled with the ideas on their own first.

The idea of proof

The theorem stated above is proved by using the results about base angles of arches, as described in the discussion of figure 2.26. This theorem incorporates the base angle results, since for the base angle of an arch, k represents the number of sides of the polygon that lie in the interior of the angle, and this number is the same number that the students used in their work with arches (see tables 2.3 and 2.4, where the first column gives the number of sides of the regular polygon that are sides of the arch). Writing a complete proof of this theorem for general n is complicated to organize and present, even though the students have proved all the important steps either in general or in examples. Students would probably not be able to follow such a complicated proof when written out with all details, so a more effective course might be to recall, with pictures, the key steps, and let some details go.

Anyone who is familiar with the inscribed angle theorem from the theory of circles can see that this is a special case of that theorem. If the polygon is inscribed in a circle with center O, the angle POQ has measure $\left(k/_n \right) \cdot 360°$, and hence angle PAQ (where A is a vertex of the polygon) measures $\left(k/_n \right) \cdot 180°$.

Further figures

This analysis can be extended to even more figures based on regular polygons. For example, figure 2.31 shows five angles with vertex A, each with measure $\left(1/_5 \right) \cdot (1 \text{ straight angle})$, or 36 degrees.

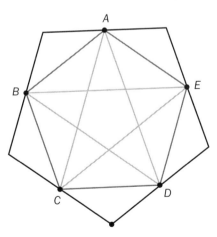

Fig. 2.31. Finding measures of angles with vertex A

Key elements and reasoning habits

The tasks suggested in Angles and Diagonals can scaffold reasoning as students move step by step from an application of familiar ideas to the development of new, more complex concepts. Table 2.5 identifies the key elements and reasoning habits (NCTM 2009, pp. 9–10, 21, 55) illustrated in this set of tasks and the accompanying classroom scenarios.

Table 2.5
Key Elements and Reasoning Habits Illustrated in Angles and Diagonals

---

Key Elements of Measurement and Geometry

Reasonableness of answers and measurements
Conjecturing about geometric objects
Reasoning inductively about shape relationships to test conjectures
Construction and evaluation of geometric arguments
Developing and evaluating deductive geometric arguments to make sense of geometric situations

---

Reasoning Habits

Analyzing a problem
Looking for hidden structure
- Finding equivalent forms of expressions that reveal different aspects of a problem
- Looking for patterns and relationships

Implementing a strategy
Organizing a solution
- Formulating an algorithm
- Using the structure of a problem

Monitoring progress toward a solution by changing or modifying strategies

Seeking and using connections
Connecting different representations

Reflecting on a solution
Considering the reasonableness of a solution
Reconciling different approaches to solving the problem
Generalizing a solution to a broader class of problems

Conclusion

The two sets of tasks presented in this chapter—the first, on the areas of polygons, and the second, on angles and diagonals of regular polygons—offer examples of the processes of exploring and making conjectures that can lead students to a deeper understanding of two-dimensional geometric concepts—and all mathematics. Although these lessons might be considered time-consuming, creating real understanding avoids the need for repetitious review and engages students. When students discover and can freely explain their discoveries to others, the question, "Why do we need to do this?" is usually not heard.

The first set of tasks, Triangles and Quadrilaterals—Exploring Area Formulas, is based on an actual classroom lesson, presented during a double block (100 minutes). The students were actively involved and participated in the development of ideas throughout the lesson. The discussions illustrate their thinking and the interest that they took in the concepts under investigation. Even when the students were later given traditional questions on a test, the teacher reported that they earned better grades on the unit involving area than previous groups of students in her experience.

The second set of tasks brings together a mix of mathematical ideas arising from an initial conjecture about angles in a regular polygon. The tasks suggest ways to take students beyond the usual discussion of interior and exterior angle measures. The investigation introduces new terms and units of measure as tools for finding patterns and justifying the original conjecture. Many reasoning habits can be developed and strengthened in this way.

Chapter 3

Reasoning about Surface Area and Volume

We live in a three-dimensional world, yet much of what children study in school geometry relates to figures in the plane. The high school mathematics curriculum should offer opportunities for students to explore their world and its connections to mathematics. Investigations of three-dimensional objects are a valuable part of this curriculum. The tasks and classroom scenarios presented in chapter 2 illustrate ways to introduce students to investigating and discovering relationships within and among two-dimensional objects. Similar approaches can be used to motivate student reasoning in three dimensions. A solid understanding of two-dimensional geometry can lead to important reasoning about relationships within and among three-dimensional objects.

The mathematical tasks and classroom discussions in this chapter focus on making sense of two central measurable features of three-dimensional objects—surface area and volume. The chapter presents two classroom investigations. The first, Covering a Sphere, takes a practical problem and shows how students can use tools and prior knowledge of two-dimensional objects to derive a method for finding the surface area of a sphere. The second investigation, Developing a Formula for Volume, consists of a set of tasks from a classroom experience that engaged students in analyzing known volume formulas and developing techniques to *discover* and begin to *prove the validity* of the formula for the volume of a pyramid. Although the students did not arrive at a formal proof, they engaged in powerful reasoning supported by their understanding of the underlying concepts.

Covering a Sphere

The first task, Covering a Sphere, poses a "real-world" problem:

How Much Canvas?

Homestead Central High School's senior class service learning project is to provide art pieces for the school. One important project is to design and build a sphere whose surface will be covered with designs inscribed with pertinent community information. The woodshop teacher has agreed to have her students assist by building the sphere, which will have a diameter of 1.25 meters. This sphere will be covered with canvas on which the art students will paint their designs. How much canvas will the students need to cover this sphere?

In the classroom

In Mrs. Perez's tenth-grade geometry classroom, students are arranged in groups of three or four. They have the statement of the problem above, several pieces of half-centimeter grid paper, scissors, pencils, rulers, tape, and a tennis ball that has been cut into two hemispheres, as shown in figure 3.1a.

(Any type of hollow ball will do for the exploration.) One of the hemispheres has been cut again into two congruent slices, as shown in figure 3.1b. The following discussion ensues.

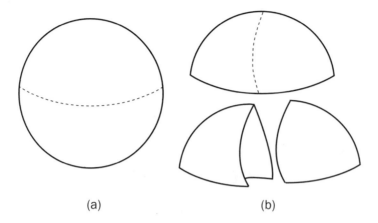

(a)   (b)

Fig. 3.1. Hollow ball cut into hemispheres and slices

| Mrs. Perez: | [*Holding an uncut tennis ball*] Let's start small. How could we determine how much grid paper we would need to cover this tennis ball? At your tables, you have several tools that might be useful. Think about the question in your groups for a minute, and then we'll share some ideas. |

[*Eva, Henry, and Maggie are working together, sharing initial thoughts. Mrs. Perez listens in on the conversation.*]

Eva:	We could wrap the grid paper around the ball. But how would we know when we had wrapped just enough? We don't want to overlap the paper too much.
Henry:	I think we should cut strips of paper and paste them on the ball.
Mrs. Perez:	What would the strips look like? How thin? Would all the strips be the same width and length?
Henry:	We would have to make the strips different lengths, wrapping the longer strips around the biggest part of the ball and then smaller strips around the rest of the ball.
Mrs. Perez:	What do you mean by the "biggest part of the ball"?
Henry:	Where the ball is the widest. Around the middle.
Maggie:	[*Examining the tennis ball in Mrs. Perez's hand*] Isn't the ball the same "width" all around? No matter how I turn the ball, it's just as wide in the middle. So how can we describe the "middle" of the ball?
Henry:	OK [*using a pencil to indicate how he would go around the middle of the ball*]. It's like the equator.

At this point, Mrs. Perez asks the class if they have come up with any ideas for finding how much grid paper they would need to cover the ball. Henry is the first to share his thoughts.

| Henry: | I thought we could use strips of the grid paper. A long strip would go around the middle, like the equator, and shorter strips could cover the rest. The strips could all be the same width—like two or three squares across. |

Mrs. Perez:	When you say "like the equator," you are describing what is called a *great circle*. What do you think the definition of *great circle* is?
Maggie:	The biggest circle that a ball has.
Mrs. Perez:	It would be the biggest circle that we could get by slicing this sphere; that's correct. If you take two halves of the tennis ball, what do you notice about the two halves?
Henry:	They are exactly the same. So, when you slice on a great circle, you get two congruent pieces.
Mrs. Perez:	That's right. But there is something else that's important to notice about a great circle. When you look at one half of the tennis ball on your table, the cut side makes a circle that is a great circle. What relationship does the center of the circle have with the center of the sphere?
Xuhui:	They're the same. The great circle contains the center of the sphere.
Mrs. Perez:	That's right. What if I cut the ball in half a different way? What would happen?
Eva:	It would be the same. You would get a great circle again, and the center would be the same as the center of the sphere. That means the radius of the great circle is the same as the radius of the sphere.

Other students agree that is the case. Thinking of a great circle as the circle obtained by slicing a sphere through its center is a clear and concise way of defining *great circle*. The discussion continues:

Mrs. Perez:	Do we have any other ideas on how to "wrap" this ball in grid paper?
Xuhui:	We could cut each little square on the grid paper and stick it to the tennis ball, then count how many we stuck on. But that would take a long time.
Henry:	Yes, but I think that would give us a closer estimate of how much paper we would need.
Mrs. Perez:	Is there any shortcut you can think of? Do we have to do the whole ball or even half the ball?
Brandon:	Oh, that's why you gave us the tennis ball already cut up. We could do just one of the quarter slices and then add up all the little squares and multiply by 4! Pretty sneaky, Mrs. P.
Mrs. Perez:	It seems that would work. Is there any relationship between this great circle that we see when we look at the hemisphere and the amount of paper that it would take to wrap the whole ball?
Maggie:	The bigger the circle, the more paper you will need!
Mrs. Perez:	So, do you think we could possibly predict how much paper we would need just from knowing the size of this great circle? [*Students indicate that they agree that this should be the case.*] How can we check that out?
Xuhui:	We could draw the great circle on the grid paper by placing the hemisphere on the paper, make several copies, cut them out, and maybe try to cover the sphere with several of these circles.
Mrs. Perez:	That sounds like an idea to investigate. Let's try to predict how many circles of grid paper we would need to cover the sphere.

Eva:	At least 2.
Xuhui:	I think it will be more like 3 or 4.
Mrs. Perez:	How about if you investigate this in your groups. For the next five minutes, let's all look at the piece of the ball that is $^1/_4$ of the sphere. Essentially, what I want you to do is think about the relationship between its outer surface area and the area of the great circle traced on your grid paper from the hemisphere.
Henry:	I think the circle's area is bigger.
Mrs. Perez:	How could you check this out? Talk with your group members.

The students return to their small groups. Some begin tracing the great circle from the hemisphere while others discuss alternate methods. Eva seems eager to share an idea with fellow group members Henry and Maggie:

Eva:	We could flatten that piece of the ball [*indicating the $^1/_4$ slice*] on the grid paper and count or estimate how many little squares are covered. Then we can compare with the area of the circle you've drawn, Henry.
Henry:	Let's do that.

Henry takes the $^1/_4$ slice of the tennis ball out of Maggie's hand and attempts to flatten and trace it on the half-centimeter grid paper next to the circle he already traced (see fig. 3.2). Maggie is leaning over, counting the number of half-centimeter squares to estimate the area of the great circle. Eva and Henry are working on estimating the area of the $^1/_4$ slice. They discover that the number of half-centimeter squares for the circle seems to be about the same as the number of these squares covered by the "flattened out" $^1/_4$ of the ball. Several other groups have come to the same conclusion by using various methods. Mrs. Perez calls for the students' attention to discuss what they have found.

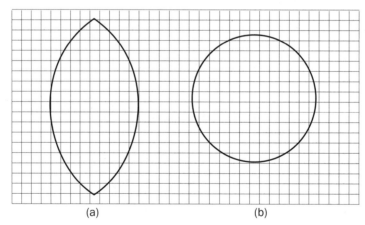

(a)   (b)

Fig. 3.2. Tracing (a) a flattened ¼ slice of a tennis ball and (b) a great circle of the ball

Activities in which students arrive at the conclusion that the area of a circle is equal to π times the squared radius can be found in several textbooks and Internet mathematics resources. It is helpful for students to do an activity such as this in class before they encounter the concept of the surface area of a sphere.

Mrs. Perez: Let's review what you've found. I noticed that most groups found that the area of the great circle is about the same as the surface area of the $^1/_4$ slice. Some groups have begun looking at the area formula for a circle to find an algebraic relationship between the areas. What did we decide the area of a circle is?

Several students: It's equal to πr^2.

Mrs. Perez: What does the r refer to?

Brandon: The r is the radius—the radius of the circle.

Mrs. Perez: And because the circle is a great circle of the sphere, what else do you know about r?

Brandon: Oh yes, we just said that r is the radius of the sphere, too.

Mrs. Perez: If what you've discovered about the relationship between the area of the great circle and the surface area of $^1/_4$ of the sphere is correct—if they are equal—how could we now express the surface area of our sphere?

Henry: Oh, by multiplying that formula, the πr^2, by 4.

[*Others nod in agreement.*]

Mrs. Perez: Let's say that our tennis ball is approximately 6 centimeters in diameter. Use what you've just shared with me to figure out how much paper we would need to cover the ball.

The students work in their groups and come up with some correct and some incorrect answers (including a result from squaring 6 centimeters—the diameter, not the radius. Finally, they agree on approximately 113.1 square centimeters—the rounded product of 4, π, and 3^2). The class discussion resumes:

Mrs. Perez: Let's go back to our original problem. How much canvas will the seniors need to cover their art sphere? Remember the diameter of the sphere for that project is 1.25 meters.

[*Students compute, obtaining approximately 4.91 square meters.*]

Mrs. Perez: Let's get back to our tennis ball. We estimated that the diameter of this tennis ball is approximately 6 cm. How can we get a better approximation of that?

The students discuss possibilities, coming to an agreement that placing the great circle of the cut tennis ball on a ruler marked in centimeters would give a good approximation. One student cautions that it is important to try to get the edge of the ruler through the center of the great circle in order to measure the diameter. Another student says that they can do this by moving the cut side of the tennis ball around on the ruler until they get the biggest measurement. Eva has a different suggestion:

Eva: Another way would be to trace the great circle on a piece of paper, wrap a piece of string around the traced circle, and get the circumference. We can measure the circumference by measuring the string, and divide it by π to get the diameter. Remember, we showed a few weeks ago that the circumference of the circle divided by diameter is just about 3.15. We talked about this ratio really being equal to π. So, diameter is just circumference divided by π.

Before working with three-dimensional objects, students should encounter activities in which they measure circumferences and diameters of circular objects of different dimensions and "discover" that the ratio of the circumference to the diameter is always "a little more than 3" and arrive at the formula for circumference (diameter times π).

Groups are asked to choose either of the two suggestions for finding a closer approximation for the diameter of the tennis ball. After a short time, the class agrees to use 6.5 cm as the diameter of a standard tennis ball (the actual diameter is 6.7 cm).

Mrs. Perez then asks the students to figure out a better approximation for the surface area of this ball. The students apply what they have found, multiplying $4 \times (\frac{1}{2} \times 6.5)^2 \times \pi$, obtaining approximately 132.7 square centimeters.

As an interesting extension to the problem of covering the sphere, students might compare the surface area of a sphere of radius r with the lateral surface area of a cylinder with a height of $2r$ and a base diameter of $2r$:

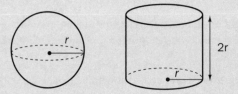

Depending on the students' algebra backgrounds, a comparison of formulas for surface area of sphere and surface area of an open cylinder would show the relationship. The sphere has a surface area of $4\pi r^2$, whereas the open cylinder is the shape of a rectangle with length $2\pi r$ (representing the circumference of the sphere) and width $2r$. Thus, the area of the rectangle is $2\pi r \times 2r$, or $4\pi r^2$. This comparison provides an algebraic proof that the areas are the same.

Discussion of students' work in covering a sphere

The geometry class that participated in this activity was a regular tenth-grade class. These students had completed one successful year of algebra. After the teacher introduced the task, the students completed it during a 55-minute period, which limited the time for further exploration. To finish the task in this block of time, students need to have had previous lessons on area and circumference of circles. Activities that help students become familiar with the terminology and the rationale for the formulas used to compute area and circumference are helpful in building understanding of three-dimensional objects.

"Discovering" formulas helps students familiarize themselves with the process that mathematicians may have gone through in the past, making sense of formulas and their relationships to objects that they describe. Having students work in groups and use objects with which they are familiar brings geometry to the realm of everyday life. In the classroom presented above, the teacher moved the students back and forth between small-group work and whole-class discussion. In this way, the teacher gave the students responsibility for *making conjectures* and for beginning to develop the *reasoning to support them*.

More specifically, the activities led students to the discovery that the surface area of a sphere is four times the area of a great circle of the sphere. Thus, they derived a useful formula for finding a solution to the original problem. At the same time, the teacher offered many ideas that helped lead the students to more substantial reasoning. She guided students by introducing (or reintroducing) geometry concepts that would help them make sense of the given situation. For example, Mrs. Perez used the term *great circle* to describe Henry's "equator." She then asked students to analyze this new concept and its relationship to the original sphere. She pushed the students to *reflect on their own reasoning* as well as the arguments offered by classmates to check on the reasonableness of their suggestions. Although the class pursued no formal justifications, the students' work helped them make connections between two-dimensional and three-dimensional objects. The activity also helped students make sense of the formula for the surface area of a sphere.

When students have finished the activity, working together to write about what they have discovered is beneficial. Doing so can solidify their understanding of why the formula works, helping them to remember it and enabling them to apply the processes to new geometric shapes. Assessment of their learning might involve having them work through the task of discovering the formula for the surface area of a cone, in the process thinking about the shapes that make up the lateral surface and the base of this shape. Table 3.1 summarizes the key elements and reasoning habits (NCTM 2009, pp. 9–10, 55) that are illustrated by the students' work in Covering a Sphere.

Table 3.1
Key Elements and Reasoning Habits Illustrated in Covering a Sphere

---

Key Elements of Reasoning and Sense Making with Geometry

Conjecturing about geometric objects

Analyzing shape relationships to formulate conjectures

Construction and evaluation of geometric arguments

Developing and evaluating deductive arguments (informal and formal) to make sense of geometric situations

---

Reasoning Habits

Analyzing a problem

Looking for hidden structure by finding equivalent forms of expressions that reveal different aspects of a problem

Implementing a strategy

Organizing the solution

- Formulating an algorithm
- Using the structure of a problem

Monitoring progress toward a solution

- Reviewing a chosen strategy
- Changing or modifying strategies as necessary

Seeking and using connections across different representations

Reflecting on a solution

Considering the reasonableness of a solution

Generalizing a solution

Developing a Formula for Volume

Mrs. Higa's ninth-grade geometry class assembles in her classroom. It is near the end of the third quarter. In the last unit, the students explored area, developing formulas for the areas of a wide range of figures by using a mix of informal observations and more formal proofs. (See chapter 2 for an example of students developing the formula for the area of a trapezoid.) This week Mrs. Higa's students have started a new unit on volume, beginning with discussions of what volume is and then developing a general approach to finding the volume of prisms. This approach involves thinking about stacking "layers" of cubic units to fill a prism, with the area of the base indicating the number of cubes in each layer and the height providing the number of layers. This work has led to an agreement on the general formula for volume of prisms:

$$V = (\text{area of the base})(\text{height}) = Bh.$$

Task 1

As the students take their seats in groups of three or four, Mrs. Higa distributes identical pyramids made out of cardstock, making sure that each student has at least one. Each pyramid has a square base and a height equal to the length of a side of the base. Mrs. Higa begins to introduce task 1: "Today we are going to start thinking about how we might find the volume of a new geometric solid. Take a look at the solids on your tables. Does anyone remember what this solid is called?"

Several students respond that it is called a *pyramid*, and one student, Jeremy, offers that it is a *square pyramid*. Mrs. Higa responds, "Great. Now I'd like you all to work together on the task on the screen":

> **Task 1:** Use your pyramids to suggest how we can find the volume of a pyramid.

(The problem in task 1 is adapted from Martin [1996].)

In the classroom

As Mrs. Higa walks around the room, she notices that a number of groups are trying to fit their pyramids together. This does not surprise her, since in the last unit the students regularly used a similar strategy to find the areas of polygons, by fitting copies of the same polygon together to form a more familiar polygon. Jenna's group is having a particularly heated discussion.

Mrs. Higa:	What seems to be the problem, guys?
Jenna:	We can't figure out how to fit them together. There keeps being a gap.
Mrs. Higa:	OK, I see what you mean. They don't fit together in a nice way to give a figure like a prism. I would recommend that you explore this approach a little more but also think about other approaches that you might take.

As Mrs. Higa walks by Keola's group, the members ask if they can have some scissors. "You want to cut up my beautiful pyramids?" Mrs. Higa asks in mock horror, while giving them a pair of scissors. "Cut up just one. I need enough for next period." Again, she does not find the students' approach surprising, since the class routinely uses what they call the "cut-and-paste" method to find area, sometimes literally cutting up (decomposing) printed or hand-drawn copies of polygons to recompose the pieces to form a more familiar polygon whose area they can find more easily.

Andre's group is trying a similar approach but has given up on the models and is discussing a sketch showing a triangle enclosed inside a rectangle, as shown in figure 3.3. Mrs. Higa engages the members of the group in the following discussion:

Mrs. Higa:	What idea are you working on?
Andre:	We couldn't get the pyramids to fit, but we think this might be like a triangle and a rectangle, so we decided to draw a picture.

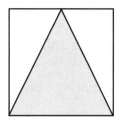

Fig. 3.3. A triangle enclosed in a
rectangle

Mrs. Higa: Where are you getting the triangle?

Andre: That's like looking at the pyramid from the side [*holds up the pyramid to illustrate*].

Mrs. Higa: And where are you getting the rectangle?

Shawna: We're thinking, just like we can fit a triangle inside of a rectangle and it is half, we should be able to fit the pyramid inside a rectangular prism.

[*The group members discuss how the pyramid would just fit into the prism.*]

Mrs. Higa: So what do you think that will tell us about the volume?

Shawna: We were thinking maybe a half, but we're not quite sure.

Mrs. Higa gives a two-minute warning for the groups to finish up what they are doing to be ready to present their findings. She makes one last pass around the room, deciding which groups to call on for different viewpoints. She is pleased with the students' attempts to form conjectures about the situation and notes that their discussions have incorporated a number of important reasoning habits related to analyzing a problem, including *identifying relevant concepts* and representations, *looking for hidden structure* as they tried to see how the pyramids can be fit together, and *applying previously learned concepts* and approaches to the new situation.

> Mrs. Higa's students are accustomed to using processes of investigation to discover mathematical ideas. This brief classroom scenario also shows that they are accustomed to expecting that they will share their reasoning with others. Although this class has clearly had "practice," reasoning and sense making can happen in any classroom where teachers ask, "How do you know?" and students have the time to think and explain.

Mrs. Higa: OK, let's get started. Let me have your attention, please. Stop what you're doing and look up here.

[*The class becomes quiet, and Mrs. Higa calls on one of the groups that were trying to fit the pyramids together. The group members hold up their solids, now taped together, to illustrate their efforts, and Karole speaks for the group.*]

Karole:	We haven't quite gotten this to work out yet. There either ends up being a gap between them or else they don't really make a nice figure.
Mrs. Higa:	[*Turning to the next group*] OK, Keola's group tried something a little different. What did you do?
Sara:	[*Speaking for the group*] We tried to cut a pyramid in half so that we could see how it would fit together.
Carlos:	[*Speaking from another group*] You cut up her pyramid? Wow!
Sara:	Yeah, it was a pain getting it to stay together, and then we couldn't really get that to work either. We were trying to make it into a cube, but it wouldn't really fit together right.

[*Mrs. Higa calls on one of the groups that were making a drawing and asks the members to share their work. One member, Rachel, shows the group's sketch, which looks like the drawing made by Andre's group; see fig. 3.3.*]

Justin:	[*Speaking for the group*] Well, we decided to just draw it, and we thought that this [*indicating the volume of the pyramid*] would be one half, just like with a triangle inside a rectangle.
Sasha:	[*Speaking from Andre's group*] Yeah, that's what we thought, too. If you look at the pyramid from the side, it's like a triangle, so that should work.
Sara:	I'm not so sure about that. When we were fitting our two halves together, it looked like it would be less than half, because it's half two different ways.
Karole:	Oh! I get it. When you look at it from the front, it's a triangle. But if you look at it from the side, it's a triangle going that way, too. So it can't be a half. I'd guess maybe a fourth?
Mrs. Higa:	[*Addressing the whole class*] So, what do we think? Is the volume of your pyramid half of a corresponding rectangular prism? A fourth? Something else?

[*Several other groups present ideas based on what they have been doing. Finally, a student named Johnny makes a suggestion.*]

Johnny:	I kind of think I saw a formula for this one time. I really can't remember it, but I think it had $^1/_3$ in it.
Mrs. Higa:	So maybe a third? How are we ever going to decide?

Through this discussion, Mrs. Higa has helped the groups *reflect on possible solutions*, including considering the different approaches taken by students in the class and *monitoring their usefulness* in reaching a solution. However, the groups are now more or less at a standstill, so some additional scaffolding is needed to help them dig deeper into the problem.

> Mrs. Higa's purposeful decision to call on students with varying ideas, rather than asking for volunteers to share ideas, leads to an examination of a range of possible solution methods. This technique can be useful in helping all students see a range of problem-solving techniques and consider the reasonableness of some approaches.

Mrs. Higa goes to her desk and pulls out a larger copy of the pyramid that the students have been using, but without the base. She also pulls out a cube with the same dimensions, also with one of its bases missing (see fig. 3.4). Then, with a flourish, she pulls out a large bag of rice. "Now does any-one have an idea?" The students show some excitement at the prospect of using the physical models to test their conjectures. Mrs. Higa asks Anne, who is usually fairly quiet, to come to the the front of the room to help.

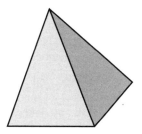
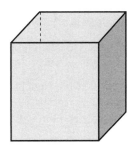

Fig. 3.4. A square pyramid and a rectangular prism with congruent bases and heights

Mrs. Higa: OK, before we get started, what do we notice about the pyramid and the prism?

[*With a little prodding, the students observe that they have the same base and the same height.*]

Mrs. Higa: Now what should we do?

Several students: Fill up the pyramid with rice, and then pour the rice into the prism!

Mrs. Higa: And then what?

Fred: Do it again until it's full!

Mrs. Higa: OK, it's time to take a stand [*seeking a show of hands*]. How many people think I'll need to fill up the pyramid 2 times to fill the prism? How many think it will be 4? Anyone for 3?

[*The students offer a mix of responses, with the majority voting for 4.*]

Mrs. Higa: OK, Anne. It's time to try it out. Pour in one pyramid-full of rice and hold up the prism so the class can see.

Keola: I still think it will be 2.

Carlos: No way! That's not nearly half!

Karole: I tell you, it's going to be 4!

[*Mrs. Higa asks Anne to put in a second pyramid-full of rice. Some students are becoming less sure of their answers, and as Anne pours in the third pyramid-full of rice, there is a general exclamation, "So it's 3!"*]

Keola: But I'm still not convinced. Is it exactly 3?

Mrs. Higa: Keola raises a good point here. Are we sure it's exactly 3? Anne, what do you think?

Anne: Well, it looks pretty close, but maybe it's not quite 3. It's not quite to the top.

[*The class offers several explanations, such as, "Maybe you didn't fill the pyramids completely full," and, "It looks like the sides of the prism are bulging a little."*]

Mrs. Higa:	So are we completely convinced? Beyond a shadow of a doubt?
Shawna:	But that's the formula Johnny remembered, so it's probably right.
Mrs. Higa:	But what have we discussed over and over this semester?

[*Class members call out a variety of responses, such as, "Looks can be deceiving," and, "We should be able to explain why things work, not just go on appearances."*]

Mrs. Higa:	Very good. So you might want to think about what it would take to really convince you that this is true, beyond the fact that Johnny remembered a formula.

Through this discussion, Mrs. Higa has reinforced a viewpoint that she has worked hard all year to instill—empirical observations are not enough to guarantee that a generalization is true. Conjectures need to be confirmed through deductive arguments built on established geometric principles.

Discussion of students' work on task 1

Mrs. Higa had many goals for her students related to reasoning and sense making. Earlier sections have highlighted these, but this section reemphasizes them, to show how Mrs. Higa's classroom choices led to student reasoning. As Mrs. Perez did in the example Covering a Sphere, Mrs. Higa chose an investigation that permitted her students to *build on previously learned concepts*. In this episode, the students' previous work in finding the volumes of prisms gave them tools and concepts that they could use in this new situation. Students explored the new figure, looking for elements in the new task that were similar to their earlier work with prisms, so that they could begin *making conjectures*.

This exploration phase was followed by a class discussion during which Mrs. Higa asked students to share initial ideas and consider the *reasonableness of these ideas* in light of the task. At this point, she did not expect an answer or even a conjecture, but she did expect students to *consider possible solution methods* carefully. As Mrs. Higa directed the investigation that followed, in which the students used rice to make an initial comparison of the pyramid's volume to the cube's volume, she continued to keep the students' ideas center stage by asking the students what they saw, what they would do, and why. At last, with a *conjecture developed*, the students recognized that this was not the final step. They understood that the conjecture had to be confirmed in a more rigorous, mathematical way. Table 3.2 summarizes the key elements and reasoning habits (NCTM 2009, pp. 9–10, 55) that are illustrated by the students' work in task 1 of Developing a Formula for Volume.

Table 3.2
Key Elements and Reasoning Habits Illustrated in Task 1

Key Elements of Reasoning and Sense Making with Geometry
> **Conjecturing about geometric objects**
>> *Reasoning inductively*
>> • Analyzing visual and/or measurable relationships
> **Geometric connections**
>> *Using geometric ideas*, including spatial visualization, in other areas of mathematics

---

Reasoning Habits
> **Analyzing a problem**
>> *Looking for hidden structure*
>> *Seeking patterns and relationships*
>> *Applying previously learned concepts*
>> *Making preliminary conjectures*

Table 3.2—*Continued*

Implementing a strategy
Selecting appropriate representations
Monitoring progress toward a solution
- Reviewing a chosen strategy
- Analyzing a problem further

Reflecting on a solution
Considering the reasonableness of a solution
Justifying or validating a solution or method

Task 2

Now that the students have explored and formed conjectures about the volume of the pyramid, Mrs. Higa distributes a sheet of problems for the class to start working on. The first problem presents task 2:

> **Task 2:** The tribal people of Enyaj decide to build a monument of marble to honor their exalted queen, Higasius. Allysonius and Shawnasium submit their design (shown in fig. 3.5). Each layer in their 10-layer monument will be one meter thick and will be centered on top of the next one. The top layer will be 1 meter by 1 meter, and the layer below it will be 2 meters by 2 meters. The length and width will continue to increase by 1 meter in each successively lower layer. Find the total volume of Queen Higasius's monument.

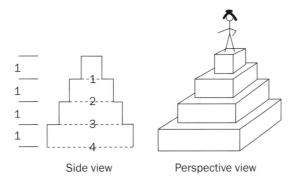

Side view   Perspective view

Fig. 3.5. Two views of the proposed monument

Additional problems on the sheet relate to the volume of prisms, helping students develop their understanding of that topic. Mrs. Higa directs the students to get started on the monument problem, working together in their groups. She urges the students not to forget an important preliminary: "Read through the whole problem and be sure that you know what it is asking you to do."

After a few minutes, Mrs. Higa asks, "So who thinks that they know what they need to do?" Several students offer their explanations, and feeling confident that the students have a good sense of what the problem is asking, she dismisses them as the bell rings: "Finish up your work on this problem, and see what you can do with the other problems on the sheet as well. Remember, explain your answers!"

In the classroom

The next day, as the students file in, Mrs. Higa reminds them that they should get into their groups and discuss what they did for homework. As they talk, she assigns each group a problem to present to the class. She begins with the presentations on the prism problems and then turns to the first problem—task 2 above. She chooses Shawna's group to present its work on the problem.

Shawna:	First, we would like to say that Mrs. Higa is totally crazy. Who is this Shawnasium? And Higasius? Very funny. Anyway, here is our work [*shows the table in fig. 3.6*]. For the top layer, we got 1 by 1 by 1, so it's just 1 cubic meter. Then the next layer is still 1 meter high, but it's 2 by 2, so that's 4. And so on, down to 10 by 10 by 1. So we got a total of 385.

Length (m)	Width (m)	Height (m)	Volume (m³)
1	1	1	1
2	2	1	4
3	3	1	9
4	4	1	16
5	5	1	25
6	6	1	36
7	7	1	49
8	8	1	64
9	9	1	81
10	10	1	100
Total			385

Fig. 3.6. Shawna's group's table to calculate the volume of the monument

Mrs. Higa:	OK, 385 what?
Shawna:	Cubic meters.
Mrs. Higa:	Does everyone agree with Shawna's group? Is 385 cubic meters right?
	[*A few groups check on their calculations, and everyone agrees with Shawna's group's response.*]
Mrs. Higa:	So how did we find the volume?
Karole:	By adding up the layers!
Mrs. Higa:	Does this sound familiar?
Several students:	That's just like we did with prisms!
Mrs. Higa:	[*Pausing a moment before asking*] So, what does this have to do with the problem we were doing yesterday?
Allyson:	Well, it looks kind of like a pyramid, except the sides aren't straight.
Mrs. Higa:	What does she mean by that?
Shawna:	The sides don't go straight up to the point.
Mrs. Higa:	Does anyone remember what we call that point?
Johnny:	The *vertex!*
Mrs. Higa:	So how does this help us answer our problem about the volume of a pyramid? If this were a true pyramid, what would you be able to say about it?

[*Students offer various observations, such as, "The base is 10 by 10," "The vertex would be at the center of the top layer," and, "The height is also 10 meters."*]

Jenna: The volume is actually too much.

Mrs. Higa: Can you explain why you think that?

Jenna: There is actually a little extra on each step. [*She draws a picture to illustrate; see fig. 3.7.*]

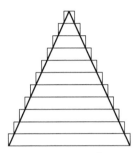

Fig. 3.7. Jenna's drawing

[*Mrs. Higa pulls out a model of a stepped pyramid that is on the same scale as the square pyramid and prism of the previous day.*]

Mrs. Higa: I just happened to have this sitting around. Could this maybe be helpful?

[*Andrew comes up and holds the stepped pyramid and square pyramid so that one is behind the other.*]

Andrew: See, this one has extra on each step, all the way up!

Mrs. Higa: Well then, if we buy that, what can we say about the volume of the real pyramid? Will this help at all in answering our question?

Jenna: Well, we know that it's less than 385, but we still don't know its exact volume.

Mrs. Higa: Suppose that our answer yesterday was that two pyramids fit into a prism with the same base and height. What would the volume of this pyramid be, without the steps?

Carlos: Um, 500, so it can't be half.

Anne: Wait. Where did that 500 come from?

Mrs. Higa: [*To Carlos*] So why don't you tell Anne again how you got 500.

[*Carlos explains that the cube is 10 meters by 10 meters by 10 meters, so that if two pyramids fitted into it, each would have to be 500 cubic meters.*]

Mrs. Higa: So do we agree that it can't be 2?

[*The students generally assent.*]

Mrs. Higa: OK, then, could it be $1/4$?

Keola: Well, that's a whole lot less, so probably not.

Mrs. Higa: So talk it over in your groups. Could it possibly be $1/4$? How could we be sure?

[The students confer in groups, generally agreeing that it wasn't $^1/_4$ but without knowing how to explain beyond saying that the 385 cubic meters for the stepped pyramid seems way too big for that.]

Mrs. Higa:	So, we're kind of thinking it should be $^1/_3$, but we'd like to have a better explanation. Does anyone remember a two-dimensional shape that we had some real difficulties with? And we couldn't find a straightforward way to find its area?
Bart:	The circle! That's kind of like this because it didn't just fit together to make a nice shape.

Earlier, when working with a circle, the class had cut a paper model into "slices" that they put together to make something resembling a parallelogram. It wasn't a perfect parallelogram, however, as shown in figure 3.8.

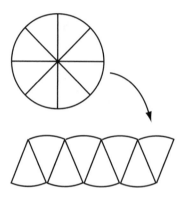

Fig. 3.8. Finding the area of a circle
with a parallelogram approximation

In that lesson, Mrs. Higa had helped the class to see that using more and more slices made the configuration of alternating slices look more and more like a parallelogram. Now she wants the class to use the same idea with the stepped pyramid to see that using more and more steps will provide a more and more precise answer. She leads the class through a review of what they did with the circle and then turns the discussion again to the pyramid:

Mrs. Higa:	So how does that idea help us with our pyramid?
Sasha:	If we made more steps, it would fit the pyramid better.
Mrs. Higa:	What does Sasha mean by that?
Sara:	Like if we used 20 steps instead of 10, it would be less bumpy.
Justin:	Look, if I used two steps instead of one, it would look like this [*shows a drawing similar to fig. 3.9 on the front board*]. So, the shaded area, the little rectangle, wouldn't be counted in. So, the two steps get closer to the slant of the pyramid, but still give more than the pyramid. Adding more steps means taking out more of the extra stuff.

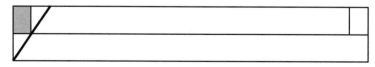

Fig. 3.9. Two rectangular steps give a closer estimate than one step.

Mrs. Higa: OK, so in your groups, see if you can figure out what the volume of the stepped pyramid would be with 20 steps. I've put your directions up on the screen.

Discussion of students' work on task 2

The second task focused on introducing the students to a possible strategy for validating their conjecture about the volume of a pyramid. As the students investigated the situation in task 2, Mrs. Higa played a central role in helping them make connections to previous material and understand relevant relationships. By the end of the class discussion, students had begun to formulate an algorithm for testing their conjecture.

Task 3

Mrs. Higa's class continues to test their algorithm for the volume of the pyramid as they work together on task 3:

> **Task 3:** Using a pyramid with the same dimensions as in task 2, find the area of a stepped pyramid that has 20 steps instead of 10 steps.

Mrs. Higa gives the groups some time to discuss and investigate solution methods for this task, and then she calls the class together to discuss the height of each layer, the dimensions of the first layer, and how the dimensions change from layer to layer. She then sends them back to work.

In the classroom

When Mrs. Higa brings the class together again, she asks for their conclusions. Allyson's group presents their solution. They have used a table much like that used by Shawna's group for a pyramid with 10 steps and conclude that the answer is 358.75 (see fig. 3.10). The other groups concur with this conclusion.

Step	Volume (cubic meters) (length × width × height)	Step	Volume (cubic meters) (length × width × height)
1	$0.5 \times 0.5 \times 0.5 = 0.125$	11	$5.5 \times 5.5 \times 0.5 = 15.125$
2	$1 \times 1 \times 0.5 = 0.500$	12	$6 \times 6 \times 0.5 = 18.000$
3	$1.5 \times 1.5 \times 0.5 = 1.125$	13	$6.5 \times 6.5 \times 0.5 = 21.125$
4	$2 \times 2 \times 0.5 = 2.000$	14	$7 \times 7 \times 0.5 = 24.500$
5	$2.5 \times 2.5 \times 0.5 = 3.125$	15	$7.5 \times 7.5 \times 0.5 = 28.125$
6	$3 \times 3 \times 0.5 = 4.500$	16	$8 \times 8 \times 0.5 = 32.000$
7	$3.5 \times 3.5 \times 0.5 = 6.125$	17	$8.5 \times 8.5 \times 0.5 = 36.125$
8	$4 \times 4 \times 0.5 = 8.000$	18	$9 \times 9 \times 0.5 = 40.500$
9	$4.5 \times 4.5 \times 0.5 = 10.125$	19	$9.5 \times 9.5 \times 0.5 = 45.125$
10	$5 \times 5 \times 0.5 = 12.500$	20	$10 \times 10 \times 0.5 = 50.000$
		Total	358.750

Fig. 3.10. A table showing the volume of the 20-step "pyramid"

Mrs. Higa then moves the students forward in the investigation by asking, "So, what if we wanted an even better answer? What might we want to do?" A student immediately responds, "How about if we used 40 steps?" As a class, the students then discuss the fact that the height of each layer will now be 0.25 meters. The length of the side of the base of the first layer should be 0.25 meters, 0.5 meters for the second layer, and so on. However, Mrs. Higa wants to make sure that the students remember to check the reasonableness of the approach they are taking:

Mrs. Higa:	What tells us that we are on the right track here?
Robert:	Well, 40 × 0.25 = 10, so that will use up the whole height. That means we've got the height of each layer figured out correctly.
Mrs. Higa:	And what about the base?
Rachel:	If we did that, the last base would be 40 × 0.25, which is 10, so that's what we want.
Mrs. Higa:	So if we use 40 steps, we'll get a better estimate for the volume of the pyramid. Then let's do it! Are you ready to get to work on this?
Keola:	[*Groaning*] There has to be an easier way. This will take forever. Can't we just fit the pyramids together somehow to show that three of them will give us a rectangular prism with the correct base area and height?
Anne:	Couldn't we just use the laptops to speed this up?

Once the students have *recognized the patterns and relationships* within the situation, Mrs. Higa sees little value in having them do all the computations. The school recently invested in a set of low-cost laptops that can be checked out for classroom use, and Mrs. Higa has been hoping that someone will make this suggestion. In fact, using a spreadsheet will help the students explore the situation more systematically. Consequently, Mrs. Higa is enthusiastic about Anne's suggestion: "That sounds like a great idea. But let's talk this through a little first."

She opens a spreadsheet on her computer and projects it on a screen for the class to see. Mrs. Higa then leads the students through the process of identifying the relevant variables that they need to set up, including the height of the layer, the dimensions of the base, the area of the base, and finally the volume of the layer.

Mrs. Higa then invites one person from each group to get a laptop from the cart and tells all the groups to try to set up a spreadsheet in the way that she just demonstrated. As she walks around the room, she encourages the students to use formulas rather than typing in the numbers individually: "That just might come in handy later!"

After most of the groups have finished, Mrs. Higa calls Rachel to the front of the room to show her group's work on the projector. Rachel connects her computer to the projector and scrolls through the numbers on the screen, showing that her group got 345.9 cubic meters as the volume of the 40-step pyramid.

Mrs. Higa reminds Rachel to elaborate, saying, "Don't just show us what you have; tell us what you did." She asks Rachel to show the various formulas that her group used and asks other class members to explain why they used those formulas. She is pleased with their ability to relate the formulas to the physical situation.

Mrs. Higa:	So, did everyone get 345.9 for their answer?
Sara:	Actually, we got a few more numbers than that, but that's close.
Mrs. Higa:	So if we agree that this is close to the answer, what will that tell us?
Marta:	Well, it's smaller than what we got for 20 steps, so maybe it will be $1/3$.

Mrs. Higa: Why do you think she is saying $1/_3$? Where is that coming from?

Carlos: Like we did before, 1000 meters cubed would be the volume of the cube, and if it is really $1/_3$, then we should be getting something close to 333 meters cubed for the volume of the pyramid.

Mrs. Higa: So do you think that is happening? Is it getting close to $1/_3$?

Sara: Maybe we should try more layers, like a thousand.

Mrs. Higa glances at the clock and decides that it is time to wrap things up. Explaining, "I just happen to have a spreadsheet set up that will compute any number of layers we want," she switches the project back to her computer and displays a spreadsheet with values calculated for a 1000-step pyramid. She shows the students that the volume calculated for the monument with 1000 steps is 333.8335 cubic meters. She remarks, "Is this looking more hopeful? We're getting closer to convincing ourselves that the volume of the pyramid is $1/_3$ the volume of the cube."

With time running out, Mrs. Higa quickly calls for questions that the students still might ask. Students call out a number of questions, and she helps to solidify several central ones:

- Will the volume of the pyramid be exactly $1/_3$ of the volume of the cube?
- Is there a way to show that the volume of a square pyramid with a base and height of 10 meters is exactly $1/_3$ the volume of a cube with the same base?
- Would the result be the same in the case of a square pyramid with base and height *h,* along with a rectangular prism with the same base and height as the pyramid?

Mrs. Higa wraps up the discussion, telling the students that at this point they don't really have the tools to answer these questions completely. "In fact," she says, "you'll have to wait for calculus to really prove how this works. But it's true that $1/_3$ does work! I know we usually would like something more definite to go on, but we'll just have to settle for this for now."

In some ways, Mrs. Higa is dissatisfied to leave things at that point, but she recognizes that there really is no good alternative, given the students' level of mathematical knowledge. For homework, Mrs. Higa gives them several problems that ask them to apply the knowledge that they have developed, and she also asks them to write a one-paragraph summary of what they have concluded from their investigations of the volume of a pyramid.

In the next class period, Mrs. Higa builds on the students' answers to the homework to derive the typical formula for the volume of a pyramid, relating it to the volume of a prism with the same base and height. She also allows a couple minutes for them to share their thoughts from the writing assignment. Although this investigation has taken some time, she feels confident that the students will not soon forget the relationship between the volumes of pyramids and prisms, and that their abilities to reason about and make sense of mathematics have been challenged and extended through the experience.

Discussion of students' work on task 3

After the class discussion of task 2, Mrs. Higa engaged the students in analyzing the problem a bit further by introducing task 3. With her questioning (e.g., "What if we want a better answer?" "Will it give a reasonable result?" and "What other questions need to be answered?"), she helped guide the students to review their chosen solution strategies and decide whether to continue. This process, in fact, led to the idea of modifying the strategy with the help of technology. To use the spreadsheet in calculating the total volume of the stepped pyramid, students were challenged to develop appropriate algebraic formulas for the geometric situation, thus connecting the content areas.

Students' mathematical reasoning came into play once more as they shared their results from the spreadsheet. They shared numerical answers and also explained how the algebraic formulas related to the physical situation. This helped all students validate the solution. Even so, these results did not provide a rigorous mathematical justification for the formula for the volume of a pyramid. Although Mrs. Higa's students did not have the mathematical background to develop a full proof, they were able to test the reasonableness of their conjecture ($V = \frac{1}{3} Bh$) through empirical trials with the stepped pyramid. Table 3.3 summarizes the key elements and reasoning habits (NCTM 2009, pp. 9–10, 21, 31) that are illustrated by the students' work in tasks 2 and 3 of Developing a Formula for Volume.

Table 3.3
Key Elements and Reasoning Habits Illustrated in Tasks 2 and 3

Key Elements of Reasoning and Sense Making with Number and Measurement and Algebra

Reasonableness of answers
Judging correct order of magnitude with appropriate units
Connecting algebra with geometry
Representing geometric situations algebraically

Reasoning Habits

Analyzing a problem
Seeking patterns and relationships
Implementing a strategy
Monitoring progress toward a solution
- Reviewing a chosen strategy
- Analyzing the problem further
Reflecting on a solution
Interpreting a solution
Justifying or validating a solution method

Conclusion

Both of this chapter's investigations, Covering a Sphere and Developing a Formula for Volume, demonstrate that students who have a solid understanding of two-dimensional geometry should be able to extend their reasoning skills to construct new knowledge (or support existing knowledge) about three-dimensional concepts and objects. These examples illustrate possible methods for structuring activities and guiding students' development of conjectures. Because of the nature of the geometric concepts explored in the activities, students encountered key elements of number and measurement, algebra, and geometry.

Another important element of both of these examples is the teacher's questioning as a means of keeping the students on track and helping them to develop reasoning habits. Both of the teachers in the classroom vignettes relied on questioning to guide their lessons, rather than direct instruction, and this approach allowed students to engage more actively in mathematical investigations and reasoning. Both teachers also used questioning to push students to rethink previously learned concepts and to make connections to the new concepts under investigation. At times, the teachers recognized the need to guide the students' thinking in a more productive or efficient direction, as when Mrs. Higa observed Andre fitting the pyramid inside a rectangular prism and asked the rest of the class to consider the connection of Andre's idea to previous work. Even so, the students were always expected to find and explain these types of connections. They shared their explanations in small group discussions and whole-class discussions, as well as in writing after the activities were completed. Thus, student reflection became part of the sense-making process.

Chapter 4

Reasoning in Geometric Modeling

Knowledge that mathematics plays a role in everyday experiences is very important. The ability to use and reason flexibly about mathematics to solve a problem is equally valuable. These two come together in mathematical modeling to solve real-world problems. When a real-world situation calls for visualizing relationships between objects (such as distances between cell phone towers or the shape and dimensions of a hallway through which a sofa needs to be moved), the ability to develop and reason with a geometric model is crucial. Introducing high school students to problem situations within a real-world context and asking them to use their mathematical skills to model and solve the problem opens new avenues for learning mathematics. When students confront such problems, they are more likely to explore a range of solution methods, reason about the strengths of a particular method, search for connections among areas of mathematics, and employ various reasoning habits, such as *monitoring progress* and *reflecting* on the viability of a solution.

This chapter presents only one mathematical investigation—an expanded version of an example in *Focus in High School Mathematics: Reasoning and Sense Making* (NCTM 2009). In that book, Example 17: Clearing the Bridge (pp. 64–70) briefly outlines a sequence of tasks related to a situation frequently encountered in the world of commercial transport: a tractor-trailer truck becomes stuck under a bridge. This chapter's extended version of the original example provides a series of tasks that begin with the basic geometric setup of the situation. The tasks then increase in difficulty as the students explore the mathematics of the situation more deeply, including related trigonometry and function concepts. Each task has its own point of closure, but this juncture also leads naturally to more questions that fuel students' interest and can carry those with sufficient understanding through subsequent tasks. Students have opportunities to *make connections* throughout the discussion. The text specifies other relevant reasoning habits at the end of each task.

In the set of tasks that follow, as in modeling problems generally, students traverse the modeling cycle as they—

- encounter a real-world situation;
- build a mathematical model of the situation that includes relevant mathematical descriptors and assumptions—some of which will be simplifying assumptions;
- derive conclusions within the mathematical model;
- interpret these results in the real-world context.

The modeling process is often repeated as the model is adjusted, additional assumptions are made, or simplifying assumptions are removed.

The Problem: Clearing the Bridge

Mathematician Henry Pollak (2004) wondered why tractor trailers often got stuck under a certain underpass when the maximum clearance—the height from the roadway to the bridge—was clearly labeled on a sign. This is the situation that this chapter's investigation explores. Here, the bridge is level and located just at the base of a descending roadway, as illustrated in figure 4.1.

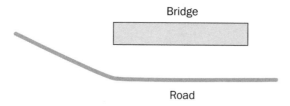

Fig. 4.1. A road descending a grade and passing under a bridge

Suppose that the truck driver knows the trailer height—that is, the distance from the top of the trailer to the road when the trailer sits on a flat road. Why might the trailer get stuck even if its height is less than the maximum clearance indicated on a sign such as that in figure 4.2? The next few tasks guide students in analyzing the situation through the use of a mathematical model.

Fig. 4.2. A sign indicating maximum clearance

Many aspects of the problem can be approached on several levels of geometric thinking. For example, task 1, which calls for an initial model of the situation, could begin with a discussion of how to build a model of a truck trailer. This would involve students in geometric thinking at van Hiele levels 1 and 2 (Burger and Shaughnessy 1986). Level 1 entails the ability to reason about shapes and other geometric configurations according to their appearance as visual wholes. At level 2, students explicitly attend to, conceptualize, and specify shapes by describing their parts and spatial relationships among the parts. So, work on the Clearing the Bridge problem could start with a level 1 activity in which students draw a picture of a truck and represent it with a rectangular box sitting on the wheels. Students can use drawings created on paper or with interactive geometry programs to develop initial visual models of the situation.

As the tasks progress, they allow more advanced students to consider the range of variables and make appropriate adjustments to enhance the model. In addition, some aspects of the problem can be used as a springboard for some new topics (e.g., asymptotes, inverse functions, or parametric equations), as well as a context for applying students' previous knowledge. Although all the tasks presented in this chapter are not appropriate for all students, the mathematics involved in the first few tasks offers a rich introduction to geometric modeling and numerous possibilities for supporting geometric reasoning. (The "Field of Vision" problem in chapter 1 offers another example of the role that geometric modeling can play in problem solving.)

It is sometimes difficult in commercially available interactive drawing programs to alter lengths or move geometric objects slightly so that measurements are whole numbers. Such was the case in selecting an example of road grade measurement in degrees for this investigation. Although the selected measurement—25.7°—could be set to round to the nearest whole number, some greater precision is desirable in work on the problem. Thus, on the one hand, this number is a reasonable one for students to use; on the other hand, a grade of 25.7° is not realistic. However, the grade has been deliberately exaggerated to make useful geometric configurations more visible. The final task involves researching and using more realistic measurements for trailer length and road grade.

Task 1: Building an initial model

As is typical in mathematical modeling, students begin the modeling process with a simplified model. In this case, they represent the trailer with a two-dimensional view that pictures the trailer box as a rectangle. For the drawing, students assume that the trailer box is attached to front and back wheels at axle level, as shown in figure 4.3.

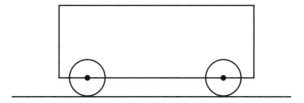

Fig. 4.3. A truck trailer attached to front and rear wheels at axle level

The students also make some assumptions about the wheels. (In this investigation, the term *wheel* includes both the rim and the tire.) First, they assume that the wheels are the same size. Second, they assume that the wheels do not flatten out on the bottom as the trailer sits on the roadway. That is, they assume that the wheels retain their circular shape while the truck sits on the roadway (see fig. 4.3). This is a simplifying assumption, since in reality tires that are correctly inflated leave a small, slightly extended footprint.

These assumptions about the trailer box and the wheels ensure that the top and bottom of the trailer box are parallel to the flat road. Having a trailer box level with the flat road is important for many reasons, including steering control, wear on tires, and load balance. However, reasoning from the assumptions to this result is not immediate, and teachers should explore it with students. Why do these assumptions lead to the fact that the top and bottom of the trailer box are parallel to the flat road?

In whatever format students encounter task 1 (class discussion, group work, or individual work), they should recognize that the assumption that the trailer wheels can be represented in the two-dimensional model by circles means that the wheels in the drawing would be tangent to the road. This means that the radius of each wheel would be perpendicular to the flat road, as indicated in figure 4.4.

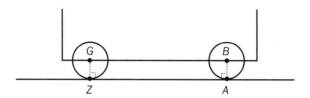

Fig. 4.4. With circles representing the wheels, the wheels
are tangent to the roadway.

Because students assume that the wheels are the same size, the radii of the wheels would be the same length. That is, in figure 4.4, $AB = ZG$. (For simplicity, this investigation denotes the length of a line segment by the segment's "unmodified" letter symbols; for example, AB, without other specification, denotes "the length of the line segment AB.") The students' attention now turns to confirming that line BG is parallel to line AZ.

In Euclidean geometry, there are many ways to prove that lines BG and AZ are parallel. Students might focus on the quadrilateral $ABGZ$, first constructing diagonal BZ, as in figure 4.5, and reasoning as below:

Fig. 4.5. Constructing $\overline{BZ}$ to aid in proving $\overline{BG} \parallel \overline{AZ}$

In addition to being equal in length, sides AB and ZG are parallel because they both are perpendicular to line AZ. Hence, the alternate interior angles GZB and ABZ are congruent. Then it is easy to see that triangles GZB and ABZ are congruent by the side-angle-side postulate for triangle congruence. Hence, $\angle ZGB$ is a right angle. Then, since segments BG and AZ are both perpendicular to line GZ, they are parallel. So, line BG is parallel to line AZ. It also follows that $\angle GBA$ is a right angle and, thus, quadrilateral $ABGZ$ is a rectangle.

Table 4.1 summarizes the key elements and reasoning habits (NCTM 2009, pp. 9–10, 55–56) that are illustrated by the students' work in task 1.

Table 4.1
Key Elements and Reasoning Habits Illustrated in Task 1

Key Elements of Reasoning and Sense Making with Geometry
    Construction and evaluation of geometric arguments
    Geometric connections and modeling

Reasoning Habits
    Analyzing a problem
        Seeking patterns and relationships
        Looking for hidden structure
    Implementing a strategy
        Making purposeful use of procedures
        Making logical deductions

Table 4.1—*Continued*

Reflecting on a solution

Justifying or validating a solution

Task 2: Identifying variables

Students now construct a two-dimensional representation of the situation that shows the danger as the trailer passes under the bridge. They make a list of the measurement items related to their representation that would be important to consider in analyzing whether or not the trailer will clear the bridge. Clearly, the height of the trailer is one, but only one, of the measurements that are important.

By reflecting and creating a visual model, students should easily see that if part of the road slopes, one set of trailer wheels is jacked up on the sloping part as the truck passes under the edge of the bridge. This causes a portion of the trailer to be raised higher than it would be on a flat surface and creates the danger that the trailer will get stuck. So the real question is, *How much higher has the trailer been raised at the point at which it passes under the bridge?* The representation that students (or groups of students) produce is likely to include at least some of the characteristics shown in the diagram in figure 4.6.

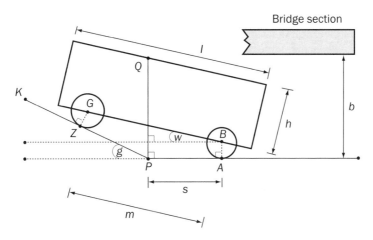

Fig. 4.6. Variables in the situation of the trailer passing under the bridge

Teachers can guide the class in reaching a consensus about a geometric model and the set of items to be measured. The tasks that follow assume that students have decided on the model in figure 4.6. Associated items to be measured include the following:

- The trailer's height, as measured from the flat ground to the top of the trailer (not labeled in fig. 4.6)

- The height, h, from the top of the trailer to the wheel axles. (This is the height of the trailer box in fig. 4.6.)

- The distance, r, from the axles to the ground. (In fig. 4.6, $r = AB$.)

- The distance, m, between the wheel axles. This distance affects the tilt of the trailer. (Students may suggest looking at the total length of the trailer, l, in addition to or in place of m at this point; task 3 addresses this issue.)

- The distance that the rightmost axle extends under the bridge. (This is denoted by s in fig. 4.6.)

- The "dangerous height" of the trailer. (This height is PQ in fig. 4.6.)

- The grade, g, of the sloping part of the road. (This is indicated by an angle measure in fig. 4.6; see the discussion of the grade of the road in the note on p. 83.)

- The tilt of the truck, w, as indicated by an angle measure in figure 4.6. (This item will play a role in some subsequent tasks. Students may or may not suggest it at this time, and the teacher can decide whether or not to discuss it now. One possibility would be to wait until task 4.)

- The height, b, of the bridge as measured from the flat part of the road to the bottom of the bridge

Discussion of students' work on task 2

For the initial modeling in task 2, some students may prefer to visualize the situation mentally or by making a few draft sketches. Others may prefer making (or using) a physical representation of a truck and a road. Still others may prefer to use an interactive drawing utility and create a representation such as the one in figure 4.8. Students can make use of such a utility in a number of tasks at an early level of geometric reasoning. For example, given both the wheel radius and the distance between axles, they might be able to figure out how to construct a drawing like that in figure 4.4. To do so, they could reason that since the quadrilateral in figure 4.5 is a rectangle, and the opposite sides of a rectangle are equal in length, the distance between the axles is the same as the distance between the points where the wheels touch the roadway. So, to construct their drawing, they might develop the following procedure:

1. Create a segment marking the axle separation along the line representing the roadway

2. Construct a perpendicular line at one end of the axle segment constructed in step 1

3. Shorten the line to a segment that represents the wheel radius

4. Construct a congruent segment at the other end of the axle separation segment, either by repeating steps 2 and 3 or by translating the segment along the axle segment

5. Connect the endpoints of the radii with a segment to represent the bottom of the trailer

Task 2 calls for a drawing that represents the danger for the truck in the situation of passing under the bridge. Not all students would represent the road with two straight line segments that meet at a vertex. This is a simplifying assumption. In reality, there would be a curving transition from the sloping portion of the road to the level portion. However, in all likelihood this portion of the road would be far away from the onset of dangerous trailer elevation. Students can discuss the effects of a gentler transition after completing their analysis of this simpler case.

Students might suggest including additional measurements, such as the distance from the flat roadway to the sloping wheel of the truck (PZ) or the maximum height of the truck. These could be included in the list. In any case, classroom discussion should make the points that students might (a) list more measurements than they actually need, or (b) discover that they have omitted some measurements that they need to add later as they go deeper into the problem. Adding measurements that turn out to be essential is simply part of adjusting a mathematical model to fit the situation.

The assumption that the trailer sits on the wheel axles may also be a point of discussion. Again, this is a simplifying assumption. Subsequent iterations of the model could consider more complicated wheel and trailer configurations, including more wheels. Other angles might be suggested as measures of tilt. An alternative choice would be $\angle ZAP$. However, angle w, as originally defined, will be used subsequently in this set of tasks.

Task 2 and several subsequent tasks are related to the theme of reasoning with number and measurement elaborated in *Focus on High School Mathematics: Reasoning and Sense Making*. The idea of *what* to measure is a central task in the analysis of this problem. For example, the issue is not the precision (or round-off error) of the clearance sign that indicates the height from the horizontal portion of the road to the bridge. Rather, the maximum length of the line segment *PQ,* as compared to the bridge height, is what is important. As students later discover, this length is not the same for every truck and, in fact, varies with a number of the parameters in the bulleted list above. Although the total height of the trailer from the level road is not easy to label in the graphic, students who give a little thought to it at this point can easily see that it is the sum of the height of the trailer box, *h,* and the axle height, *AB.*

A second measurement decision involves making a reasonable choice of measurement units. In the real-world context of the problem, this issue arises in the decision about how to measure the grade (slope) of the road. Road grade is a measure of the road incline from the horizontal. In this sequence of activities, the road grade, *g,* is the degree measure of the acute angle formed by the sloping road segment with the extended horizontal road segment. In typical situations, road grade is measured as a percentage, as in a 9% grade (see fig. 4.7). That is, the typical grade measurement identified on highway signs is equivalent to tan (*g*) converted to a percentage.

Fig. 4.7. The road grade is typically given as a percentage.

The choice of unit of measure for the road grade is up to the teacher. However, the selection of degrees for the measure of grade angle provides consistency with how students would measure other angles in the geometric context of the problem. In particular, students compare the tilt angle, *w,* and the grade angle, *g,* in a later activity. So the tasks use degrees throughout the investigation. It is worth stating the (obvious) assumptions in the model that the grade of the road, *g,* is always nonnegative and will never be as great as 90°. The teacher might lead a discussion of grade as a percentage measure at this point or soon afterward.

A central goal of the analysis in subsequent tasks is to uncover how components of the model interact with each other. Although versions of these tasks can be implemented without technology, an interactive visual model can bring the mathematical inquiry to life and help to reveal several aspects of the effects of changes in parameters on the "dangerous height." As presented, the tasks assume that students have available an interactive model such as that captured statically in figure 4.8.

Using an interactive model such as the one depicted in the figure, students can adjust the distance, *m,* between the axles by moving a slider. The effect is to draw the elevated truck wheel up the incline without moving the wheel that is on the level surface. The size of the wheels can also be adjusted with a slider that changes axle height, *r,* from the road. Moving this slider adjusts both wheels equally. Without changing the trailer's dimensions, students can move it further along the level portion of the road (in either direction) by dragging the point *A* with the mouse. They can adjust the grade, *g,* of the sloping portion of the road by dragging the point *K.* Any of these four adjustments changes the tilt angle, *w.* Moving a slider adjusts the height, *h,* of the trailer box by raising the roof of the box. A slider for trailer overhang, *v* (not shown explicitly in fig. 4.8), lengthens the trailer extension beyond the wheels. (In the interactive model pictured, this adjustment affects both wheel overhangs equally.)

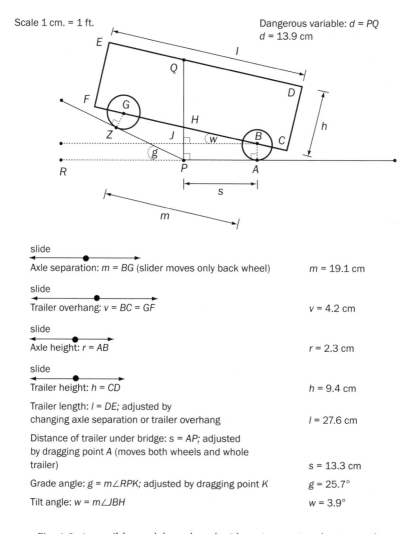

Fig. 4.8. A possible model produced with an interactive drawing utility

Through exploration of the model, students notice that the dangerous height is unaffected by changes in one of these parameters, trailer overhang (*v*), though it is altered in various ways by changes in others. Tasks 3–5 progress through different levels of analysis that investigate the change in the dangerous height, *d,* which results from a change in the distance, *m,* between the axles. Table 4.2 summarizes the key elements and reasoning habits (NCTM 2009, pp. 9–10, 55–56) that are illustrated by the students' work in task 2.

Table 4.2
Key Elements and Reasoning Habits Illustrated in Task 2

Key Elements of Reasoning and Sense Making with Geometry
   Conjecturing about geometric objects
      Analyzing configurations
      Making conjectures about relationships
   Geometric connections and modeling

Table 4.2—*Continued*

Reasoning habits

    Analyzing a problem

       Seeking patterns and relationships

       Looking for hidden structure

Task 3: Investigating "dangerous height"

It is fairly obvious that raising the trailer height, by increasing either the height, h, of the trailer box or the axle height by using bigger wheels, will have a fairly direct effect on the dangerous height, d. Students can explore changes in these variables by using an interactive model. However, Pollak noted that the length of the truck might also be responsible for putting the truck in danger of getting stuck. He thought that this length might be undervalued as a contributing factor in some cases. Task 3 turns students' attention to this suggestion. The interactive model allows them to explore a relationship between the length, l, of the trailer and the dangerous height, d, either by increasing m, the distance between the axles (by using the axle separation slider), or increasing the distance, v, that the trailer extends beyond each axle using (by using the "trailer overhang" slider). Students work through parts (a) and (b) of task 3:

    a.   Explore what happens to the dangerous height, d, as you alter the size of m, v, or both.

    b.   Can you make any conjecture relating to a pattern in the way d changes as you alter m, v, or both? For any relationship between the variable d and any of the three variables m, v, and l, can you create a representation that helps you formulate any conjectures about the way that d changes? Explain.

In the Classroom

Students working on part (a) of task 3 discover that the overhang, v, of the trailer has no effect on d, its dangerous height. What follows is a hypothetical class discussion and sample classroom interaction between teacher and students related to part (b):

Teacher: Let's share some of your answers to the question in part (b) about relationships among the variables, particularly with respect to the "dangerous height" of the trailer.

Jake: Using the interactive model and sliding the m slider, our group made a table that shows that the dangerous variable, d, increases as the length of the trailer increases [*shows a table like that in fig. 4.9*]. In the table, m goes up one unit at a time but the change in d varies. We saw a pattern in the changes in the d values. They go 0.4, 0.4, 0.4, 0.3, 0.2, 0.3, 0.2, 0.2, 0.2, 0.2, 0.1, 0.2, 0.1, 0.1, 0.1. The numbers bounce around a little bit, but we think round-off error could play a role here. With some bumps, the change in d gets smaller. We aren't confident enough to try to make a formula to fit the data that we have so far. But we think the change in d is going to continue to become less and less.

Teacher: That's a good start. Did anyone try something different?

Alisa: Our group used the drawing program to plot a graph of d as a function of m, and my graph also shows that d goes up when m does. [*Shows a graph like that in fig. 4.10.*] We couldn't find a formula, either. The graph looks pretty straight, but if you look closely, it looks like it tapers off when the length of the trailer gets larger… See? [*Traces the graph with her finger.*] We don't think it will ever get past 20.

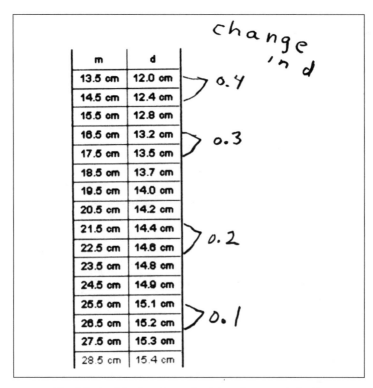

Fig. 4.9. A table showing paired data for *m* and *d* with students' calculations for change in *d*

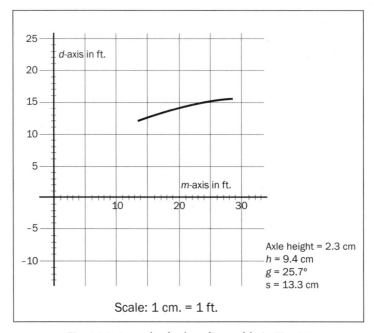

Fig. 4.10. A graph of values from table in Fig. 4.9

Teacher:	Do you think it's linear?
Oriana:	If it were linear, it couldn't taper off.
Jake:	Wait… Look at our table… If it were linear, I could take any two points from the table and get the same slope… But since our first coordinate change is always 1, our changes can be interpreted as change in the second coordinate divided by change in the first coordinate. That is, the numbers we got are all slopes. If the graph were linear, all the slopes would be the same. But we got different slopes, like 0.4, 0.3, 0.2, and 0.1. Our round-off error couldn't account for this big a difference. I say it's not linear!
Teacher:	That's a convincing argument about this model.

In some classrooms, teachers might take this opportunity to introduce the concept of *horizontal asymptotes*. Regardless of whether the students give the concept a name here or have encountered it prior to this series of tasks, some discussion such as the one suggested below can help them formulate a conjecture.

The teacher might ask, "So, what do you think would happen to the graph of *d,* the dangerous height, as a function of *m,* the axle separation, if we continued to increase *m*? Would *d* have to stop increasing? Could *d* continue to increase but not pass 20 feet? What do you think?" After some class discussion, the class might be ready to conjecture that the graph can increase indefinitely but not reach the value 20. If so, the teacher might proceed with the next set of questions:

- If the graph can increase indefinitely but never reach 20, what might the graph look like?
- Do you think there might be a "lowest" barrier that it would never reach—say, some number a little lower than 20?
- If so, what would we mean by a "lowest barrier"?

However, the class might not be ready to make this conjecture. Students might form a different conjecture, such as that the graph has to stop increasing. Or the class might not be confident enough to make a conjecture on the basis of the information that they have and might instead choose to leave this as a continuing question. The next task—task 4—provides an opportunity for students to probe such conjectures or resolve continuing questions. Whatever the case, as part of the next step in the modeling paradigm, students should interpret the impact of their conjecture in the real-world context of the trailer passing under the bridge.

Discussion of students' work on task 3

The teacher in the scenario above used students' work to begin digging deeper into the situation. She referred to Jake's table and Alisa's graph and asked all students to consider the meaning of these results in relation to the original context. This is a good time for teachers to highlight the connections between geometry and algebra.

Using a table and a graph helped the students draw some mathematical conclusions and raise questions. Moreover, working with the interactive model enabled them to produce a graph in a way that is different from using a symbolic formula. In general, most students would not be able to construct a symbolic representation of a relationship between *m* and *d* at this point. Discussing any strategies that students used in trying to find such a relationship would be worthwhile. Task 4 will suggest one approach to a symbolic relationship as students move back into the mathematical model. The result will be useful in providing information about the real-world context. Task 5 will extend this approach. Table 4.3 summarizes the key elements and reasoning habits (NCTM 2009, pp. 9–10, 21, 41, 55) that are illustrated by the students' work in task 3.

Table 4.3
Key Elements and Reasoning Habits Illustrated in Task 3

Key Elements of Reasoning and Sense Making with Geometry, Functions, and Number and Measurement

Conjecturing about geometric objects
Multiple representations of functions
Using families of functions as mathematical models
Approximation and error

Reasoning habits

Analyzing a problem
Seeking patterns and relationships
Looking for hidden structure
Applying previously learned concepts
Making preliminary deductions and conjectures
Implementing a strategy
Making purposeful use of procedures
Making logical deductions based on progress so far
Monitoring progress toward a solution
• Reviewing a chosen strategy
Seeking and using connections
Connecting different representations
Reflecting on a solution
Interpreting a solution

Task 4: Making connections with algebra and trigonometry

Task 4 consists of four parts, (a)–(d). Students should complete the first part before seeing the second, since a portion of the answer to the question in the first part is evident from the wording of the second. (Alternatively, the teacher could change the wording of the second part.)

Part (a)

Students can complete the first part of task 4 either as homework or in class on the day before encountering the second part. Their task in the first part—part (a)—is as follows:

> **Part (a):** We have found it useful to look at a table of values and the graph of the dangerous variable, d, as a function of m, the axle separation. To move toward a symbolic representation of a relationship between these two variables, let's analyze the problem a little further. Using the interactive model, we noticed that the size of d changed when you change the size of m. Does anything else change?

In the classroom

Students may need to explore by using the interactive program before answering the question in part (a). The drawing has more than one characteristic that changes. For example, by virtue of the way in which the program works, the location of the wheel on the slope changes. However, the intent of the question in part (a) is to induce students to observe that the tilt angle, w, changes. That is, an increase in m results in an increase in w. Conversely, it is also valuable for students to notice that if wheel size, lower wheel position, and road grade stay the same, an increase in the tilt angle would *require* moving the wheel on the slope up the slope to keep both wheels of the truck on the roadway. Thus, m would *need* to increase to balance an increase in w. (In this sense, m changes if and only if w does.)

Part (b)

Having noticed that an increase in *m*, the separation of the axles, produces an increase in *d*, the dangerous height of the trailer, students are ready to move on to part (b) of task 4:

> **Part (b):** In light of our results in part (a), our strategy will now be to focus on a different relationship besides that between *m* and *d*. Because of the strong relationship that we saw between *m* and *w*, let's look for a relationship between *w* and *d* instead. Can you find a symbolic relationship between *w* and *d* that doesn't make direct reference to *m*? You can use some of the other parameters that we listed in task 2, but the relationship that you represent in writing should not include *m*. In other words, by using some or all of your measurement variables, together with geometric and trigonometric concepts and results, can you find a symbolic relationship between the trailer's tilt, as measured by *w*, and the measure of the variable *d* for its dangerous height?

In the classroom

The teacher arranges the students in pairs to think about how to begin looking for a relationship between *d* and *w*. She allows the pairs to work for a short time and then brings the students together for a class discussion:

Teacher: All right, any ideas?

Pair 1: We didn't see a whole relationship, but looking at the diagram [*see fig. 4.11*] made us think that *w* is related to the part of *d* right in front of it: segment *HJ*. We then thought about what would relate that segment to *w*, and we saw that we were inside a right triangle—triangle *JBH*—so we could use trigonometry. We got that $\tan w = \dfrac{HJ}{BJ}$. But *BJ* = *s*, the distance the wheel is under the bridge. So $(s)(\tan(w)) = HJ$. That's it so far.

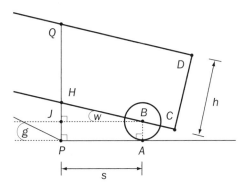

Fig. 4.11. A diagram to help in reasoning about a
relationship between *w* and *d* (*PQ*)

Teacher: That's good. That's a piece of *d*. This suggests a strategy of getting *d* in pieces. Does anyone else have ideas that focus on another piece of *d*?

Pair 2: Well, we were thinking about what happens when *w* goes up and down, and we noticed that the wheel *under* the bridge stays still and *AB*, or *r*, doesn't change. But *AB* = *PJ*, because *ABJP* is a rectangle. So, that part of *d* doesn't change. It is unaffected by *w*. Does that mean that there is no relationship between *w* and the segment *PJ*?

Teacher:	Some people would say that what you noticed could be considered a type of relationship. The relationship is that a change in *w* does not change *PJ*. In mathematics, this relationship is a constant function $f(w) = PJ$ over the values that *w* can take on. Have you seen constant functions before?
Joe:	Yes, we graphed some while we were studying linear functions. Their graphs are the ones that look like horizontal lines.
Teacher:	Good. But in this relationship, we are interested only in the values of *w* that make sense in the context that we're looking at. It is worth noting which values these are. Can anyone tell me which values of *w* make sense, here?
Arrel:	Well, when I look at our model drawing, it looks to me like *w* will never get beyond *g*. If the truck is tilted, the most it can be tilted is when the front wheel [*points to wheel under the bridge*] just hits the flat part of the road and the rest of the trailer is on the slope. Then, the trailer bottom will be parallel to the sloping road. So, the lines [*rays*] that form angle *w* will be parallel to the lines [*rays*] that make *g*. So then *w* will equal *g*.
Teacher:	Good. Looks like you are thinking that the truck is coming down the hill and going under the bridge. Let's use that idea and name the wheel on the flat road the "front wheel" and the wheel on the slope the "back wheel." That description is shorter. Can we say any more about the values of *w*?
Bethany:	Well, using similar reasoning, we'd get the other end of the range when the back wheel just touches the flat roadway. Then *w* is 0. And *w* won't go below 0.
Robert:	So, should we include *g* and 0 as possible values for *w* or not? Shouldn't our model just consider when the back wheel is really on the slope and the front wheel is really under the bridge—not just at the bottom of the hill? So, *w* will never get to 0 or get as big as *g*.

Some discussion follows, with students taking various positions about whether or not *w* should take on the values 0, *g,* or both, without reaching a consensus. After a short time, the teacher brings the discussion to closure:

Teacher:	OK, we have decided that 0 and *g* represent the extremes of the situation. Let's proceed with the situation where the angle *w* is greater than 0 but less than *g* and then see if both our reasoning and our results can extend to these extreme cases. Let's return to our question about the relationship between *w* and *d*. So far, we have that $HJ = (s)(\tan(w))$, and *PJ* stays constant as *w* changes. What's left?
Henry:	*HQ*.
Teacher:	Did any group look at that length? [*Waits until students indicate that no one got any result.*] OK then. Since we are so close, let's work on that now in groups… [*Students work for a short time with little success.*] All right. Let's all talk about this together. Let's see if there is anything we can say about *HQ*. Do you think it is longer or shorter than the height of the trailer? [*Students quickly reach a consensus that it is longer.*] Can you show me? Can you *prove* it?

[*Students start working again and continue for some minutes.*]

Sharon:	[*Speaking from group 1*] Our group can prove it! [*Shows the model in fig. 4.12.*] Look, we drew in the height of the trailer starting at the top, *Q*, and going down to the bottom of the trailer to *T*. Since we are measuring the height from *Q* to the

bottom line, we want segment QT to be perpendicular to line BH. Now, $QT = h$. But triangle HTQ is a right triangle, and segment QH is the hypotenuse, which we know is the longest side. So, $h < QH$.

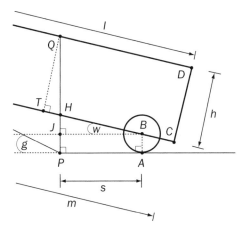

Fig. 4.12. Adding QT to aid in proving that HQ is longer than h

Teacher: That's really good, Sharon! Now we want to know how much bigger than h QH is. Sharon has added another line to our model. Let's go back to our groups and see if that extra line helps us figure out relationships involving segment QH.

[*A short time passes. Then the teacher asks for progress, and group 2 reports.*]

Mitchell: [*Speaking from group 2*] I think we've got something. Well, the first thing we noticed was that triangles BJH and QTH are similar [*Shows a drawing like that in fig. 4.12.*] See, the vertical angles at point H are equal, and both triangles have a right angle. So, their third angles have to be equal. That means that the measure of $\angle TQH$ equals w [*demonstrates on diagram and then continues*]. We looked at tan w in triangle TQH like we did in triangle JBH when we were looking at HJ, but that strategy didn't really get us far because it just gave us another new length to look at: HT. Then we switched and looked at

$$\cos w = \frac{TQ}{HQ} = \frac{h}{HQ}.$$

This means that

$$HQ = \frac{h}{\cos(w)}.$$

So, we've got a relationship between w and HQ, but it has h in it.

Teacher: That's OK. We said you could use other parameters that we had talked about. That's it! That's the missing piece! Each of you, put the pieces together and summarize the relationship.

[*Students again work for a short time before the teacher resumes.*]

Teacher: Ron, what do we have?

Ron: We've got that

$$PJ + HJ + HQ = AB + (s)(\tan(w)) + \frac{h}{\cos(w)}.$$

No, wait. I want to write this as

$$PJ + JH + HQ = AB + (s)(\tan(w)) + \frac{h}{\cos(w)}$$

because then anyone can read this and see that the three segments on the left line up to give d. So,

$$d = r + (s)(\tan(w)) + \frac{h}{\cos(w)}.$$

It is helpful to write the expression

$$d = r + (s)(\tan(w)) + \frac{h}{\cos(w)}$$

first with the letter-named segments because then students can see that they are adding three pieces, one for each segment of d. After rewriting with d, students can see that d is a function of w.

Teacher: That is really good. We looked at a table of data and got some information. We looked at a graph of data that confirmed our thoughts suggested by the data about how d might increase as m does. Tomorrow, we will analyze what this symbolic representation begins to tell us.

Part (c)

The next day, the teacher presents part (c) of task 4:

Part (c): In part (b), you found that the dangerous height, d, of the trailer has a relationship with w, its tilt angle. Does the relationship that you found work when there is zero tilt—that is, when the trailer is on a level road? Explain.

In the classroom

Students work on the task in groups. After some time, the teacher asks the members of group 4 to present their findings.

Marsha: [*Speaking for group 4*] We first noticed that if the trailer were on the flat road, the total trailer height would be just the radius, r, of the wheel plus the height of the trailer, or $r + h$. Our formula from part (b) has $\frac{h}{\cos(w)}$ in it instead of just h. But this is because, with the tilt, we are interested in QH instead of QT, and

$$\frac{h}{\cos(w)} = QH.$$

What we did next was to use the dynamic sketch to slowly lower the grade to 0, while keeping s and r the same. When the grade went down to 0, we saw three things. First, we saw that segment QT moved to fit perfectly on top of segment QH—this is why $\dfrac{h}{\cos(w)}$ is replaced by h in the flat-road formula. Next, we saw that segment HJ disappeared—this is why the term $(s)(\tan w)$ disappears in the flat-road formula. Then we saw that w went to 0.

Teacher: Group 4's reasoning makes sense. On a flat road, the truck should have no tilt. We can use the observation about w being equal to 0 when the grade is 0 (for a flat road) in our formula from part (b) to get

$$d = r + \frac{h}{\cos(0)} + (s)(\tan(0)) = r + \frac{h}{1} + (s)(0) = r + h.$$

This presentation from the group and the teacher's reformulation show that the formula for a tilting truck is a generalization of the case where the grade is 0. With more advanced students, the relationship between w and m can be discussed in terms of one-to-one functions and inverse functions. However, that is beyond the scope of the current discussion.

Part (d)

Work on task 4 continues with part (d):

Part (d): Recall that in part (b) you found a relationship between d, the dangerous height of the trailer, and w, the angle of tilt. Does that relationship help to answer the questions in task 3 about how high the variable d can go? Explain.

In the classroom

As the students enter the classroom, the teacher pairs them off for work on part (d). The discussion begins, referring to a model like that in figure 4.13.

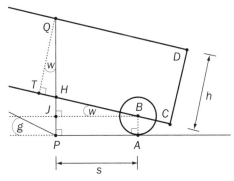

Fig. 4.13. Focusing on the relationship between w, the tilt angle, and $PQ = d$, the dangerous height of the trailer

Teacher: Does anyone remember the questions that we raised but didn't resolve in task 3?

Alex: Yeah! We wondered if d could ever go beyond 20.

July: And we were trying to decide if d could increase forever and still not go beyond 20.

Benny: And we wondered if d did increase forever, if there is a lowest number it wouldn't go beyond.

Teacher: So, what should we do now?

[*Students think on their own for a few minutes without saying anything.*]

Teacher: OK, as you often do, share some of your thoughts with your partner. [*After a few minutes, calls for the students' attention.*] Any good ideas surface?

Robert: I think Carin has some ideas.

Teacher: Carin, let's hear some of what you have been thinking.

Carin: Well, I've been looking at our formula for d and trying to figure out how big it can get... OK, r doesn't get any bigger, but $(s)(\tan(w))$ will.

Teacher: Explain what you mean when you say that $(s)(\tan(w))$ will increase.

Carin: Well, if you start to drag the wheel up the slope to make w increase, $\tan w$ will increase.

Teacher: Does $\tan w$ always increase?

Carin: Yes. I mean, no. I mean it will here because we know w will never get larger than 25.7°.

Teacher: Why 25.7°?

Carin: It's the value of the grade, g, which we happened to work with [*referring to the modeling assumption illustrated in fig. 4.13*].

Teacher: So, explain a little more about why $\tan w$ will increase.

Carin: Well, I know what the tangent function looks like for values between 0 and 25.7°. Since 25.7° is less than 90°, the tangent increases for these values of w [*see fig. 4.14*].

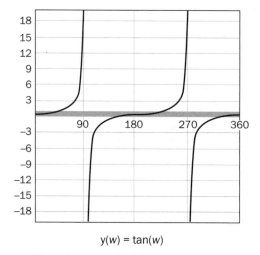

y(w) = tan(w)

Fig. 4.14. A graph of the tangent funtion

Teacher: So, would Carin's conclusion hold as long as our the grade, g, is less than 90°? [*After a short pause, the class agrees.*] Let's think back to our model. Did we make that assumption? Is it reasonable?

The students look back at the model and recall a discussion about making this assumption and discuss the idea that one could probably change the upper bound on the grade to a much lower value. But they agree to leave the upper bound at 90°.

Teacher: Do you want to continue, Carin?

Carin: OK. Well, then the term $(s)\tan(w)$ will never get any bigger than $(s)\tan(25.7°)$. I guess that, in general, what we discussed means that the term $(s)\tan(w)$ will never get bigger than $(s)\tan(g)$. So, looking at our formula,

$$d = r + (s)(\tan(w)) + \frac{h}{\cos(w)},$$

we know that the first two terms will not get bigger than $r + (s)(\tan(g))$. In the case that we were looking at, this is $2.3 + (13.3)(\tan(25.7°))$. That's as far as I got.

Teacher: Let's see if we can follow Carin's lead and finish. What do we need to do now?

Riva: We need to see how big $\dfrac{h}{\cos(w)}$ can get.

Teacher: How about if you get back into your pairs and work on this.

[*Students work for about ten minutes, and then the teacher calls for their attention.*]

Teacher: Any pairs make any progress? Jamie and Jill—what about you?

Jamie: I think we've got an answer. We did what Carin did and looked at the trig part, $\cos w$. For $0 < w < 25.7°$ we looked at the graph of the cosine [*shows a graph like that in fig. 4.15*]. As w gets bigger, the cosine gets smaller. That will make $\dfrac{h}{\cos(w)}$ get bigger. But then it can't get bigger than $\dfrac{h}{\cos(25.7°)}$. Or I guess that in general $\dfrac{h}{\cos(w)}$ won't get bigger than $\dfrac{h}{\cos(g)}$, because g is less than 90°.

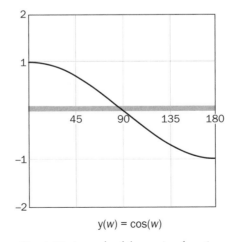

$$y(w) = \cos(w)$$

Fig. 4.15. A graph of the cosine function

Teacher: Great. Now, everyone—write a summary of our conclusion.

[After a few minutes, the teacher calls on Michael to present a summary.]

Michael: Well, I said that we had this formula for the dangerous variable, *d:*

$$d = r + (s)\tan(w)) + \frac{h}{\cos(w)}.$$

For the case that we had some data on, we showed that

$$d \le 2.3 + (13.3)(\tan(25.7°)) + \frac{9.4}{\cos(25.7°)}.$$

Or

$$d \le r + (s)(\tan(g)) + \frac{h}{\cos(g)},$$

because we are assuming $0 < g < 90°$.

Teacher: Does everyone agree? *[Waits until several students nod in agreement before continuing.]* What does this mean in the context of the truck trailer?

Sarah: Well, the first part of what we were talking about means that *QH* and *HJ* both increase as *w* increases. But it then means that the whole dangerous height, *PJ + JH + HQ,* never gets any bigger than

$$2.3 + (13.3)(\tan(25.7°)) + \frac{9.4}{\cos(25.7°)},$$

using the numbers that we were using. Or it means that the dangerous height doesn't get any bigger than

$$r + (s)(\tan(g)) + \frac{h}{\cos(g)}.$$

Teacher: Does everyone agree?

Benny: Well, we have to remember that we are keeping *r, s* and *h* fixed in what Sarah said about the general case.

Seth: Whoa, using our values for the wheel radius and the trailer box height, I just calculated

$$2.3 + (13.3)(\tan(25.7°)) + \frac{9.4}{\cos(25.7°)}.$$

It comes out to about 19.13. We guessed that the dangerous height wouldn't get past 20. We did pretty good!

Teacher: So, what about our questions of the dangerous height increasing forever?

Jake: Yeah, our argument says that *d* will increase forever…

Kate: *[Interrupting]* But it will stop when *w* gets to 25.7°.

Teacher: That was an interruption, Kate. Let's let Jake finish.

Jake: I was going to say that it increases forever but never gets above 19.13 in the case we are considering. I know 19.13 is only an approximation of the actual value.

Teacher: Now, we should consider Kate's comment. What do you think, class?

Harold: When I was using the interactive drawing program for task 3, I never could get w all the way to 25.7. I don't think you can ever get there. No matter how long m is, the top side of angle w will never be parallel to the sloping road. So, the angle will always be less than 25.7°. So, d will never stop increasing.

Teacher: Class, do you agree that we have provided a convincing argument for the conclusion that as long as w increases but doesn't reach 25.7°, d will increase but never get past

$$2.3 + (13.3)(\tan(25.7°)) + \frac{9.4}{\cos(25.7°)},$$

or approximately 19.13? [*Waits until the class agrees.*] Does it also seem that

$$2.3 + (13.3)(\tan(25.7°)) + \frac{9.4}{\cos(25.7°)}$$

is probably the lowest ceiling we could guarantee for any length truck? [*Waits again until the class agrees.*] Well, at least our arguments seem to give us some confidence in this result. OK, it seems that our success in getting a symbolic relationship between w and d did help us a resolve a major conjecture from task 3!

Discussion of students' work on task 4

In the class discussion of part (d) of task 4, the dialogue that began with Carin summarizing the fact that d increases as a function of w is an illustration of reasoning and sense making with *number* and *functions*. At the same time, Carin also used the geometric representation to help justify her conclusion. This discussion could have been replaced by an analysis of the graph of d as a function of w (see fig. 4.16).

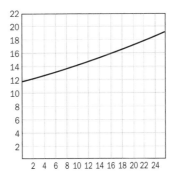

Fig. 4.16. A graph of d, the dangerous height of the trailer, as a function of w, its tilt angle

Although the graph helps in measuring how much d increases as w increases, teachers are encouraged to have students explain why d increases, thus stimulating the kind of insight and number sense that Carin and other students subsequently manifested. The discussion of the "lowest number that d doesn't go beyond" relates to the concept of *least upper bound* that students will encounter in more advanced study of the real number system. Table 4.4 summarizes the key elements and reasoning habits (NCTM 2009, pp. 9–10, 21, 31, 41, 56) that are illustrated by the students' work in task 4.

Table 4.4
Key Elements and Reasoning Habits Illustrated in Task 4

Key Elements of Reasoning and Sense Making with Geometry, Algebraic Symbols, Number and Measurement, and Functions
Geometric connections and modeling
Meaningful use of symbols
Reasoned solving
Connecting algebra and geometry
Number system properties
Multiple representations of functions
Using families of functions as mathematical models

Reasoning Habits
Analyzing a problem
Applying previously learned concepts
Reflecting on a solution
Interpreting a solution
Considering the reasonableness of a solution
Justifying or validating a solution
Generalizing a solution

The Clearing the Bridge investigation could end with task 4 for some students. However, more advanced students are on the verge of unraveling the relationship between m and d and, in so doing, they will gain results and tools to analyze the problem from additional perspectives.

Task 5: Another symbolic representation

Task 5 brings the investigation full circle as students develop a symbolic representation of the relationship between m and d that they can use to generate the graph of d as a function of m previously produced by the interactive drawing program in task 3. The symbolic representation also highlights the relevance of the grade of the road, the distance that the trailer is under the bridge, and the height of the trailer box.

At this point, students have developed a good deal of insight about the dangerous height, d, in relation to the tilt angle, w, and, to some degree, the axle separation m. They understand that w and m are related—they can see this relationship as they change the length m by using an interactive drawing program. However, questions remain about the relationship between w and m—for example, how large does m have to become before the height gets close to 19.13 feet? To answer this question, students need to examine the relationship between w and m more closely. They can use some of the techniques that they found in task 4 to help them identify such a relationship:

> Find and analyze a symbolic relationship between m, the distance between the trailer axles, and w, the tilt angle of the trailer.

Students are given this problem and a copy of the drawing in figure 4.17 for task 5. They can proceed by using either the drawing or a dynamic construction.

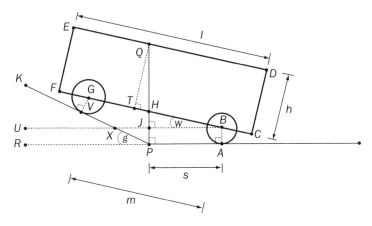

Fig. 4.17. A drawing for the students to work with in task 5

In the classroom

The students begin to work in new groups. After a while, the teacher brings the groups together to report on their progress.

Teacher: All right, how about if the members of group 6 talk about their progress.

Marion: [*Speaking from group 6*] We tried the same strategy as before. We noticed that $m = BG$. So, we tried to break up segment BG into two segments: BH and HG. We got that

$$\cos(w) = \frac{BJ}{BH} = \frac{s}{BH}.$$

So,

$$BH = \frac{s}{\cos(w)}.$$

But we didn't know what to do about HG. We were looking for another right triangle to work with, but we didn't find one—at least, not yet.

Teacher: [*Addressing the rest of the class*] Any suggestions?

Ron: You could construct a right triangle by extending line BF to meet line PK.

Teacher: Can we guarantee that these lines meet?

[*After a few moments a student, Sharon, responds.*]

Sharon: I think I can do it. Well, since lines BU and AR are parallel, the measure of $\angle UXK$ is g. So, the measure of $\angle KXB$ is $180° - g$. I'm looking at the lines PK and BF, cut by the transversal BU. Now, the sum of the measures of $\angle UXK$ and $\angle FBU$ is $(180° - g) + w$, which is less than $180°$, because $w < g$. So, by Euclid's parallel postulate, lines PK and BF intersect.

Teacher: Great! Let's call the point of intersection Z and add it to our sketch [*labels the point as in fig. 4.18*].

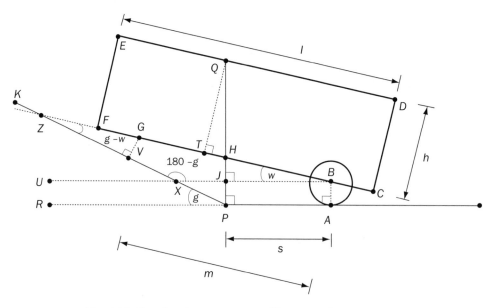

Fig. 4.18. Showing the intersection of line BF and PK as point Z

Teacher: Let's work on this in our groups to see if we can get any further. [*Allows some time to pass.*] Any progress?

Sharon: [*Speaking for group 4*] We couldn't see what relation the small right triangle ZVG had to BG. So we are trying out another strategy. This one doesn't divide segment BG into pieces—it uses all of segment BG, and then some. So, here it is: We used the law of sines with triangle ZXB because we know all the angles now. We get

$$\frac{BZ}{\sin(180 - g)} = \frac{BX}{\sin(g - w)}.$$

Now, $BZ = BG + GZ = m + GZ$. Also,

$$\sin(g - w)\frac{GV}{GZ} = \frac{r}{GZ}.$$

So,

$$GZ = \frac{r}{\sin(g - w)}.$$

That means

$$BZ = m + \frac{r}{\sin(g - w)},$$

and

$$\frac{m + \dfrac{r}{\sin(g - w)}}{\sin(180 - g)} = \frac{BX}{\sin(g - w)}.$$

That's as far as we got.

Teacher: Did other groups try different approaches?

Several groups offer strategies. For example, some students tried using triangle *PJX*. In the process, they found that $JX = JP \cdot \tan(90 - g) = (r)\tan(90 - g)$. But now this approach—as well as others—has stalled. A suggestion gets the students moving again:

Teacher: So, let's see what we all can do next by using Sharon's description of what group 4 has done so far. Each group should decide what to do and then carry out the strategy. [*Permits some time to pass.*] All right, what progress have we made? Sharon, what did your group decide to do next?

Sharon: We decided to look at *BX* because it is in our formula. We noticed that $BX = BJ + JX$. But we know that $BJ = s$, and from what some other groups found, $JX = (r)\tan(90 - g)$. So, $BX = s + (r)\tan(90 - g)$. So, now our complete formula becomes

$$\frac{m + \dfrac{r}{\sin(g - w)}}{\sin(180 - g)} = \frac{s + (r)\tan(90 - g)}{\sin(g - w)}.$$

We solved for *m* and got

$$m = \frac{\sin(180 - g)(s + (r)\tan(90 - g)) - r}{\sin(g - w)}.$$

Teacher: Wonderful! Are there other approaches?

Some groups that obtained results used slightly different geometric reasoning based on drawing auxiliary lines, such as the perpendicular line through point *X* to line *AR*. In place of the expression $(r)\tan(90 - g)$ in the final result, these groups derived the equivalent expression $\dfrac{r}{\tan(g)}$ for *JX*.

After the groups share their reasoning and results, the teacher brings the discussion back to the case at hand:

Teacher: Good. So, what would the equation look like with our set of numbers?

Ron: [*Goes to the board; speaks while writing*]

$$m = \frac{\sin(154.3)(13.3 + (2.3)\tan(64.3)) - 2.3}{\sin(25.7 - w)}.$$

Teacher: Good job, Ron! Jill, did you want to say something?

Jill: I was just going to say that this equation for *m* as a function of *w* looks like it should. I mean, thinking back to what we did in analyzing *d*, when *w* gets closer and closer to 25.7°, the difference 25.7° − *w* will get closer and closer to 0, so *m* will get bigger and bigger, like it should: The closer *w* is to 25.7°, the bigger *m* will be.

Teacher: Harold, you have been very eager to say something. What did you want to add?

Harold: I just think *m* is going to get way too big for real life when *w* gets close to 25.7°! Can we take a look at that now?

Teacher: What would we look at?

Harold: The graph of m as a function of w.

Teacher: Do you want to enter the function in the graphing calculator so that we can all see it on the view screen?

Harold: Sure! [*Enters the function in the calculator and tries graphing it, adjusting the window several times before he gets an image he likes—the graph in fig. 4.19.*]

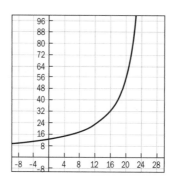

$$m(w) = \frac{\sin(154.3)(13.3 + (2.3)\tan(64.3)) - 2.3}{\sin(25.7 - w)}$$

Fig. 4.19. Harold's graph

Teacher: OK, show us your graph.

Harold: See! Look, when w is 20°, m is up over 50 feet. And when w is 22°, it has jumped to over 60 feet. And that's only the distance between the axles. We have overhang above the rear wheel of another 4.2 feet and the length of the truck cab. I think that's pretty big.

Teacher: What do you think, class?

Russell: I don't know. I've seen some pretty big trucks.

Teacher: Why don't we research that fact this evening and see what we can find out.

During the next class session, the teacher summarizes the students' work on the Clearing the Bridge investigation up to this point. The discussion of a truck trailer getting stuck under a bridge led the class to investigate how the "dangerous height," d, of a truck trailer was influenced by the distance, m, between the trailer's front and rear axles. The students looked at some data and a graphical representation generated by an interactive drawing program and showing d as a function of m. An analysis of that information allowed students to develop some conjectures about an upper limit that the dangerous height would not exceed. For the numbers that they were using in the drawing program, that upper limit was approximately 19.13 feet. They found that limit by developing a symbolic expression for the dangerous height, d, that allowed them to represent d as a function of the tilt angle, w. That insight emerged from a discussion of the relationship between the tilt angle, w, and the distance between the trailer axles, m. The students then found a way to represent that relationship between w and m symbolically. They discovered that in general terms those two relationships are

$$d(w) = r + (s)(\tan(w)) + \frac{h}{\cos(w)},$$

and

$$m(w) = \frac{\sin(180 - g)(s + (r)\tan(90 - g)) - r}{\sin(g - w)}.$$

Continuing her recap for the students, the teacher notes that they have discovered that these relationships, taken together, are dependent on several other parameters: the radius of the wheels, r; the height of the trailer box, h; and the grade of the sloping part of the road, g. For the values that the students have been using for these parameters, their functions are

$$d(w) = 2.3 + (13.3)(\tan(w)) + \frac{9.4}{\cos(w)},$$

and

$$m(w) = \frac{\sin(154.3)(13.3 + (2.3)\tan(64.3)) - 2.3}{\sin(25.7 - w)}.$$

The teacher reminds the students that ultimately they have been searching for some symbolic way to represent the functional relationship between m and d. Among other things, this would give another way to graph the ordered pairs (m, d), where d is the dangerous height corresponding to length m, the separation of the axles of the trailer. What the teacher tells the students next surprises them: "But we have that information!" She continues:

> For each value of w where $0 \le w < 25.7°$, the ordered pair $(m(w), d(w))$ tells us which m-value goes with each d-value. That is, we let w do the pairing for us. If we just graph the resulting pairings (m, d), the work of w will be behind the scenes. But we will get a view of the graph of d as a function of m from the symbolic representations that we have developed. The relationship $(m(w), d(w))$ is called a *parametric representation* between m and d with parameter w. Equations like those that we have found for m and d are called *parametric equations.*

Many common software programs and most graphing calculators have the capability of graphing parametric representations. The teacher explains the process for entering the parametric equations and setting up the calculator to display the graph:

Teacher:　　　　Here is what our example looks like [*shows the graph in fig. 4.20*]. What do you think?

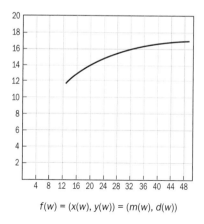

$f(w) = (x(w), y(w)) = (m(w), d(w))$

Fig. 4.20. The graph of the ordered pairs $(m(w), d(w))$

Jackie:	It matches the graph that our group drew by using the interactive drawing program when we didn't have any kind of formula! See, the dangerous height reaches about 15 feet when the trailer length gets to be about 25 feet. That matches the table, too. But this graph can go on farther without having to drag anything! Wow!
Teacher:	The fact that our symbolic representation matches the graph from the interactive program is an exciting check on our work, I agree. But other than having a more expanded version of the graph, do we get additional advantages from having our symbolic representations?
Taylor:	I think it's easier and faster to change the numbers for r, g, h, and s in the equations by looking at what happens to the graph on the calculator than it is to drag the parts in the drawing program and redraw a graph. But the drawing program really helped me understand what was happening.
Efrim:	Yeah, but there's something I've been really wanting to do that we haven't done. I've wanted to watch what happens to the dangerous height as we drag the trailer under the bridge. I mean, let's fix all the parts of the trailer and the road—set r, m, h, and g—and move s. I could do that a little with the drawing program, but I also noticed that our formula [*reads off the equation on the board*],

$$\frac{m + \dfrac{r}{\sin(g - w)}}{\sin(180 - g)} = \frac{\left(s + (r)\tan(90 - g)\right)}{\sin(g - w)},$$

	could easily be solved for s the way we solved it for m. Then we could graph (s, d) the way we graphed (m, d).
Teacher:	Efrim, that's a great extension to this problem, and we can discuss who might be interested in pursuing it tomorrow. But right now class is about to end, and I have an assignment to give everyone. As Efrim mentioned, we can now change some of the parameters that we have been using to more realistic sizes. You already did a little research into how long truck trailers might be. Do some more research and see what you can find out about road grades. How big might they be? How about axle heights—or the distance from the road to the pivot height of the trailer box? What would be a good overall height for the trailer? You can also look for some signs representing bridge clearances. See you tomorrow.

Discussion of task 5

Some adjustments and assumptions were made in the presentation of this task and the accompanying figures. First, note that the circle representing the rear wheel was omitted from figure 4.18 for clarity, and labels were added for points U and V. Second, note that point V could fall between points X and P. In a justification of the equations for m, it would be important for the teacher to mention that this case has to be considered. Although this discussion does not appear in the above classroom discussion, the equation is still valid.

Solving the "law of sines equation" for m could be accomplished by hand or with a computer algebra system (CAS). With the symbolic representation of $m(w)$ accessible, this would be an ideal time to graph the inverse function. Indeed, it is more natural to think of w as a function of m—but this involves inverse functions. This approach would be helpful in analyzing a "realistic" truck (see the Conclusion below). Whether or not the classroom investigation of Clearing the Bridge includes such a discussion depends on either the students' prior knowledge of this concept or the teacher's inclination to discuss inverse functions at this point. Table 4.5 summarizes the key elements and reasoning habits (NCTM 2009, pp. 9–10, 31, 41, 55–56) that are illustrated by the students' work in task 5.

Table 4.5
Key Elements and Reasoning Habits Illustrated in Task 5

Key Elements of Reasoning and Sense Making with Geometry, Algebraic Symbols, and Functions

    Construction and evaluation of geometric arguments
    Geometric connections and modeling
    Meaningful use of symbols
    Mindful manipulation
    Reasoned solving
    Linking expressions and functions
    Multiple representations of functions
    Using families of functions and mathematical models

Reasoning Habits

    Implementing a strategy
        Making purposeful use of procedures
        Organizing the solution
        Monitoring progress toward a solution
        • Reviewing a chosen strategy
        • Analyzing the problem further
        Making logical deductions
    Reflecting on a solution
        Considering the reasonableness of a solution
        Justifying or validating a solution

Conclusion

A class that has undertaken the Clearing the Bridge investigation might wrap up its work in many ways. A possible scenario follows:

Students bring in a wealth of information about actual truck trailers—including the fact that some trucks are built so drivers can mechanically move rear axles closer to the front axles in order to reduce the measure m that the students have been studying! The students compile a collection of values for more realistic parameters:

Grade: 9%	Although some roads have grades over 13%, 9% is fairly common. The students translate a 9% grade into an angle measure of 5.14° by tracing the tangent (as a function of degrees) graph on a calculator and determining that tan(5.14°) ≈ 0.09. So, $g = 5.14°$ is reasonable.
r: 2.8 ft.	On some trucks, the real pivot point is above the wheel axle, and this height is often about 2.8 feet.
h: 9.2 ft.	An overall trailer height of 12 feet is reasonable.
m: 48 ft.	A measure of 48 feet is within bounds of some current trailer measurements.

With these sample "realistic" measurements, the students need to figure out w. They assume that s equals 13.3 feet. On the same set of axes, they then graph the function

$$m(w) = \frac{\sin(180 - 5.14)(13.3 + (2.8)\tan(90 - 5.14)) - 2.8}{\sin(5.14 - w)}$$

and the horizontal line $y(w) = 48$ (see fig. 4.21). They determine that the w-coordinate of the point of intersection is approximately 3.7°, so for the more "realistic" measurements, $w = 3.7°$.

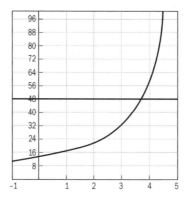

Fig. 4.21. The length of m as a function of w

Next, the students determine the dangerous height, d, for these values. They enter all the values of the parameters and the value 3.7° for the independent variable into their symbolic rule for the function d. They determine that $d\,(3.7°) \approx 12'\,8^3/_8''$. They then compare this with the bridge clearance that they settled on: *12′ 6″. They immediately note that their "realistic" trailer would hit the bridge!* One student evaluates the situation as follows:

> I think being just a couple of inches high would be worse than being a foot too high. The truck driver would probably have looked and seen that his truck was way too high if it was a whole foot off. He might not have seen the danger if he was moving and it was just a couple of inches off!

The class also determines that $d(5.14) \approx 13'\,2^3/_8''$. That would be the "safe" clearance height, analogous to the height of 19.13 feet determined earlier in the investigation. Students suggest that it would be nice if posted clearance signs included leeway. (Some, but not all, posted signs actually do include some leeway; the sign in figure 4.22 probably understates the clearance by a small amount.)

Fig. 4.22. A truck prepares to pass under a bridge with a posted clearance of 12′ 7″

However, a question remains: Is 13.3 feet—that is, the distance s that the rightmost axle is under the bridge, on the flat road—the point at which the dangerous height reaches its maximum for the students' "realistic" truck? Maybe it's time for the students to go back to work! Or, better yet, maybe it's time for your students to take their turn in exploring the problem.

The process of developing a solution to the Clearing the Bridge problem hints at many ideas that students may subsequently encounter in a calculus course. These include *limits* (e.g., as w gets closer to 25.7, $d(w)$ approaches 19.13 approximately) and *optimization*. Continued study of the mathematics related to the "Clearing the Bridge" context could occur in a calculus class.

Epilogue

The chapters in this volume have provided a range of examples of how practicing teachers have found opportunities to emphasize reasoning and sense making in geometry. Many of the contexts presented in this volume relate to topics found in traditional high school geometry curricula. The topics or concepts themselves do not guarantee that student reasoning and sense making will occur in the classroom. Rather, the emphasis on reasoning and sense making arises from the nature of the tasks and the way that classroom activities are structured around them.

Reasoning and sense making reflect the problem-solving processes that students use, as well as the ways in which they connect mathematical concepts. Asking students to share and verbalize these processes in classroom discussions helps them organize their own thinking and build on the ideas of others. The lessons outlined in this volume provide just a few examples of student reasoning. These examples also highlight teachers' decisions about when to follow a student's line of reasoning (which may be faulty) and when to provide a more solid direction to lead the reasoning in a specific direction.

We invite you to reflect on the sample lessons presented in the book and consider the ways in which reasoning habits develop and key elements of mathematics unfold in them. In most cases, students' sense making progresses through stages of understanding as they move through the activities, from making sense of new and emerging ideas to explaining or justifying a conjecture or problem solution. You might consider the teacher's role in facilitating the growth of students' reasoning. You might focus on how students' problem-solving strategies contribute to their reasoning abilities. The questions below are intended to help you reflect on reasoning and sense making in geometry. Some questions focus on stages of students' reasoning. Others focus on making it happen in the high school classroom.

Questions for Reflecting on Reasoning and Sense Making in Geometry

1. How can discussion of a task or a mathematical concept capture students' emerging ideas and enable students to share these ideas with others?

2. What is the teacher's role in facilitating the sharing of ideas and in judging students' contributions?

3. What are students' roles in sharing ideas and judging the contributions of their classmates?

4. How can the teacher modify traditional textbook problems or exercises to elicit student reasoning?

5. What can the teacher do to help students organize or reorganize ideas to promote sense making?

6. What kinds of tasks motivate students to conjecture and explain?

7. How do students analyze and make decisions about what ideas or strategies are valuable and productive in problem solving?

8. How can the teacher guide and strengthen the students' decision making?

9. What can the teacher do to promote the organization and communication of students' reasoning?

10. When is it appropriate for the teacher to require a more formal proof? What is the most effective way for students to learn how to develop a proof?

Classroom change does not occur in an instant. Identifying when and how to provide opportunities for reasoning and sense making is a first step. Implementing a plan for regularly providing these opportunities comes next. Awareness of students' strengths and limitations is valuable but may come with time.

Although we have not answered all ten questions that we posed above, we hope that the tasks, classroom episodes, and samples of students' reasoning that we have offered in this book inspire teachers to provide opportunities for their students to develop mathematical reasoning habits through the learning of geometry.

Appendix

NCTM Standards and Expectations for Grades 9–12

Geometry
Standard

Instructional programs from prekindergarten through grade 12 should enable all students to—	**Grades 9–12** **Expectations** **In grades 9–12 all students should—**
Analyze characteristics and properties of two- and three-dimensional geometric shapes and develop mathematical arguments about geometric relationships	• analyze properties and determine attributes of two- and three-dimensional objects; • explore relationships (including congruence and similarity) among classes of two- and three-dimensional geometric objects, make and test conjectures about them, and solve problems involving them; • establish the validity of geometric conjectures using deduction, prove theorems, and critique arguments made by others; • use trigonometric relationships to determine lengths and angle measures.
Specify locations and describe spatial relationships using coordinate geometry and other representational systems	• use Cartesian coordinates and other coordinate systems, such as navigational, polar, or spherical systems, to analyze geometric situations; • investigate conjectures and solve problems involving two- and three-dimensional objects represented with Cartesian coordinates.
Apply transformations and use symmetry to analyze mathematical situations	• understand and represent translations, reflections, rotations, and dilations of objects in the plane by using sketches, coordinates, vectors, function notation, and matrices; • use various representations to help understand the effects of simple transformations and their compositions.
Use visualization, spatial reasoning, and geometric modeling to solve problems	• draw and construct representations of two- and three-dimensional geometric objects using a variety of tools; • visualize three-dimensional objects from different perspectives and analyze their cross sections; • use vertex-edge graphs to model and solve problems; • use geometric models to gain insights into, and answer questions in, other areas of mathematics; • use geometric ideas to solve problems in, and gain insights into, other disciplines and other areas of interest such as art and architecture.

Number and Operations
Standard

Instructional programs from prekindergarten through grade 12 should enable all students to—	**Grades 9–12** **Expectations** **In grades 9–12 all students should—**
Understand numbers, ways of representing numbers, relationships among numbers, and number systems	• develop a deeper understanding of very large and very small numbers and of various representations of them; • compare and contrast the properties of numbers and number systems, including the rational and real numbers, and understand complex numbers as solutions to quadratic equations that do not have real solutions; • understand vectors and matrices as systems that have some of the properties of the real-number system; • use number-theory arguments to justify relationships involving whole numbers.
Understand meanings of operations and how they relate to one another	• judge the effects of such operations as multiplication, division, and computing powers and roots on the magnitudes of quantities; • develop an understanding of properties of, and representations for, the addition and multiplication of vectors and matrices; • develop an understanding of permutations and combinations as counting techniques.
Compute fluently and make reasonable estimates	• develop fluency in operations with real numbers, vectors, and matrices, using mental computation or paper-and-pencil calculations for simple cases and technology for more complicated cases; • judge the reasonableness of numerical computations and their results.

Measurement
Standard

Instructional programs from prekindergarten through grade 12 should enable all students to—	**Grades 9–12** **Expectations** **In grades 9–12 all students should—**
Understand measurable attributes of objects and the units, systems, and processes of measurement	• make decisions about units and scales that are appropriate for problem situations involving measurement.
Apply appropriate techniques, tools, and formulas to determine measurements	• analyze precision, accuracy, and approximate error in measurement situations; • understand and use formulas for the area, surface area, and volume of geometric figures, including cones, spheres, and cylinders; • apply informal concepts of successive approximation, upper and lower bounds, and limit in measurement situations; • use unit analysis to check measurement computations.

References

Battista, Michael T. "The Development of Geometric and Spatial Thinking." In *Second Handbook of Research on Mathematics Teaching and Learning,* edited by Frank K. Lester, pp. 843–908. Charlotte, N.C.: Information Age; Reston, Va.: National Council of Teachers of Mathematics, 2007.

Burger, William F., and J. Michael Shaughnessy. "Characterizing the van Hiele Levels of Development in Geometry." *Journal for Research in Mathematics Education* 17 (January 1986): 31–48.

Clements, Douglas H., and Michael T. Battista. "Geometry and Spatial Reasoning." In *Handbook of Research on Mathematics Teaching and Learning,* edited by Douglas A. Grouws, pp. 420–64. New York: Macmillan; Reston, Va.: National Council of Teachers of Mathematics, 1992.

Cooney, Thomas J., Stephen I. Brown, John A. Dossey, Georg Schrage, and Erich Ch. Wittmann. *Mathematics, Pedagogy, and Secondary Teacher Education.* Portsmouth, N.H.: Heinemann, 1999.

Hershkowitz, Rina, Tommy Dreyfus, Dani Ben-Zvi, Alex Friedlander, Nurit Hadas, Tzippora Resnick, Michal Tabach, and Baruch Schwarz. "Mathematics Curriculum Development for Computerized Environments: A Designer-Researcher-Teacher-Learner Activity." In *Handbook of International Research in Mathematics Education,* edited by Lyn D. English, pp. 657–94. Mahwah, N.J.: Erlbaum, 2002.

Martin, W. Gary. "Supporting Secondary School Students' Construction of Geometric Knowledge." In *New Directions in Research in Geometry,* edited by Annette R. Baturo, pp. 74–79. Brisbane, Australia: Queensland University of Technology, 1996.

National Council of Teachers of Mathematics (NCTM). *Curriculum and Evaluation Standards for School Mathematics.* Reston, Va.: NCTM, 1989.

———. *Principles and Standards for School Mathematics.* Reston, Va.: NCTM, 2000.

———. *Curriculum Focal Points for Prekindergarten through Grade 8 Mathematics: A Quest for Coherence.* Reston, Va.: NCTM, 2006.

———. *Mathematics Teaching Today.* 2nd ed. Updated, revised version, edited by Tami S. Martin, of *Professional Standards for Teaching Mathematics* (1991). Reston, Va.: NCTM, 2007.

———. *Focus in High School Mathematics: Reasoning and Sense Making.* Reston, Va.: NCTM, 2009.

Pollak, Henry. "Why Does a Truck So Often Get Stuck in Our Overpass?" *Consortium: Newsletter of the Consortium for Mathematics and Its Applications,* Spring–Summer 2004, 3-4.

Thompson, Patrick W. "Imagery and the Development of Mathematical Reasoning." In *Theories of Mathematical Learning,* edited by Leslie P. Steffe, Pearla Nesher, Paul Cobb, Gerald A. Goldin, and Brian Greer, pp. 267–83. Mahwah, N.J.: Lawrence Erlbaum Associates, 1996.